STEP BY STEP
GARDENING

Series Editor
ALAN TITCHMARSH

Plant Propagation

Jane Courtier

Ward Lock Limited · London

First Published in Great Britain in 1985
by Ward Lock Limited, 82 Gower Street,
London WC1E 6EQ, a Pentos Company.

House editor Denis Ingram
Designed by Bob Swan

Text filmset in Monophoto Plantin 110
by MS Filmsetting Limited,
Frome, Somerset

Printed and bound in Italy
by New Interlitho Sp A, Milan

**British Library Cataloguing in
Publication Data**

Courtier, Jane
Plant propagation.—(Ward Lock step-by-step
gardening series)
1. Plant propagation
I. Title
635′.043 SB119

ISBN 0-7063-6236-5

CONTENTS

ACKNOWLEDGEMENTS

All the line drawings are by Nils Solberg.

The drawings on pp. 8 & 9; 10; 14(2); 17(2); 26(2); 27; 28(3); 29(3); 30(2); 31; 32(3); 33(3); 34(2); 35(2); 36; 37(2); 38; 39(3); 40(2); 41(2); 44(2); 45(2); 46(2); 48(3); 50; 51(2); 52(3); 54(3); 56(2); 57(2); 58(3); 59(3); 60(3); 61(3); 62(3); 63(3); 64(2); 65(3); 66(2); 78(2); 79(3); 80(2); 81(3); 83(2); 84(3); 87; 88(3); 89(3); 91(2); & 92(2) are after the drawings on pp.21; 22; 26,25; 31,33; 67,63; 70; 71; 80; 80; 81; 81; 84; 82; 62; 65; 73; 73; 39,33; 39; 40; 55; 55; 55; 57; 52; 56; 56; 6,8,9; 92; 92; 92,91; 93; 93; 93,86; 86,87; 87; 87; 75,90; 90; 50; 50,83; 44; 44; 45; 47; 48; 48; 49; 52; & 85 respectively in *The Royal Horticultural Society's Encyclopaedia of Practical Gardening: Plant Propagation*, P. McM. Browse, Mitchell Beazley, 1979.

The drawings on pp.15; 16; 17; 18; 24(2); 25(3); 27; 30; 31; 41; 57; 69(3); & 70 are after the illustrations on pp. 32; 32; 22; 22; 45,50; 42,49,48; 60; 69; 69; 79; 109; 111; & 111,112 respectively in *The Hamlyn Guide to Plant Propagation*, Eds. S. Mitchell and B. Haynes, Hamlyn, 1969.

The drawings on pp. 10(2); 15; 16; 19(2); 22; 47; 74(2); 75(2); & 83 are after the illustrations on pp. 31,30; 21; 21; 26; 10; 48; 56; 54; & 53 respectively in *The Humex Book of Propagation*, J. Harris, Macdonald, 1980. The drawings on pp. 11; 12; 19; & 78 are after the drawings on pp. 25; 25; 33; & 40 respectively in *Gardening Through the Year*, A. Hellyer, Hamlyn, 1981. The drawings on pp. 64; 68(2); 71; & 72 are after the drawings on pp. 125; 176; 179; 188; and 188 respectively in *The Grafter's Handbook*, R.J. Garner, Faber and Faber, 1949.

The drawings on p. 6(2) are after the drawings on pp. 17 and 44 in *The Boots Guide to Greenhouse Gardening*, A. Titchmarsh, Hamlyn, 1980. The drawings on p. 73(3) are after the drawings on p.111 in *Propagation*, A. Toogood, J.M. Dent, 1980. The drawings on pp. 11, 13, 14, 20, 24, 50, 51, 67, & 72 are after illustrations the copyright of which belongs to *Amateur Gardening* magazine. The drawing (sprinkler) on p.13 is after an illustration in Hozelock Ltd. catalogue, 1983.

Publishers Note

Measurements are generally cited in imperial followed by the metric equivalent in parentheses. In a few instances, owing to pressure on space, the metric equivalent has been omitted.

FOREWORD
BY ALAN TITCHMARSH

Growing plants from scratch is one of the most satisfying things on earth. There are thrills to be had from persuading dry seeds to burst through the surface of moist soil; from encouraging a small shoot to produce roots and grow into a large plant, or maybe even a tree; from chopping up a large begonia leaf and convincing each postage-stamp-sized piece that if it really tries it can grow into a new plant that possesses all the vigour and beauty of its parent.

The obvious reason for propagating plants is to increase their numbers and save yourself money. But there are many gardeners who propagate simply for the fun of it – giving away their surplus plants once the job has been done.

If propagation seems to you to be shrouded in mystery, Jane Courtier makes it clear as crystal with detailed explanations and step-by-step diagrams. You can discover how to rejuvenate ageing house plants; how to get the best out of dust-fine seeds, and even how to grow your own rose bushes or fruit trees by budding and grafting.

Gardening is never all beer and skittles, and you'll have your failures when it comes to propagating plants. The successes, though, are what you will remember, and however sophisticated your horticultural tastes may become, nothing will ever dull that sense of achievement that comes from growing a huge plant from a tiny seed.

Equipment for seed raising: for details see text opposite.

1
SEED

Equipment

Raising plants from seed is one of the simplest, most natural and most magical methods of propagation. Much of the equipment you need is basic gardening tackle.

Any seed not sown direct into the open ground will need some form of container. Plastic seed trays and pots in a range of sizes (*a*) are convenient to use and easily cleaned. Shallow containers avoid wasting compost and make over watering less likely. Any container should have plenty of drainage holes in the base.

Compost (*b*) – and this means a specially prepared growing medium, not garden compost from the compost heap – must be of good quality to give the seedlings the best start. Proprietary brands are sterilized to kill weed seeds and pest and disease organisms: they are far superior to garden soil, which is totally unsuitable for plants growing in containers. Choose between loam-based types (such as John Innes), which look much like soil and peat-based or soilless composts, which are clean and lightweight but quickly run out of plant foods. Composts are formulated for sowing and cuttings, potting, or general purpose use.

The surface of the compost must be firm and level before sowing, and a presser (*c*) is easily made from a block of wood cut to the same size as a seed tray. Fix another small piece of wood on top as a handle.

Watering is an important operation in seed raising: fragile seedlings are easily damaged by large drops of water. Choose a small, easily handled can (*d*) fitted with a fine rose to break the water into a fountain of tiny droplets.

Most seeds need covering with a fine layer of compost after sowing, and a small sieve (*e*) makes it easier to get a fine, even covering without lumps and bumps.

Seeds in general need warmth and moisture for good germination, and there's plenty of special equipment to provide this. A rigid plastic pot cover (*f*) is one of the simplest and cheapest items. Much more sophisticated is an electric propagator (*g*), with thermostatically controlled 'bottom heat' that can be adjusted to exactly the right temperature for a particular type of seed. A plastic cover keeps in the moisture, providing a humid atmosphere, while ventilators enable you to control the humidity to prevent fungus diseases from finding it too cosy.

The main problem most avid seed sowers find is where to put all the trays. Every windowsill in the house is often pressed into use, but windowsills don't provide the best conditions. If you are prepared to delay some of your sowings until after the risk of frosts, the trays can remain in a sheltered spot outside, but better protection is given by a garden frame (*h*). Purpose-built frames can be quite pricey, but the nearest demolition site should provide sufficient old windows, bricks and timber for a handyman to knock one into shape.

Of course, as your interest in gardening grows, you will soon start hankering after a greenhouse (*i*). It's the perfect place for seed raising, and opens up many more exciting opportunities for growing a whole new range of plants – well worth saving up for! Plastic is sometimes used as a substitute for glass in greenhouses – it's certainly cheaper, but glass is by far the best material in the long run.

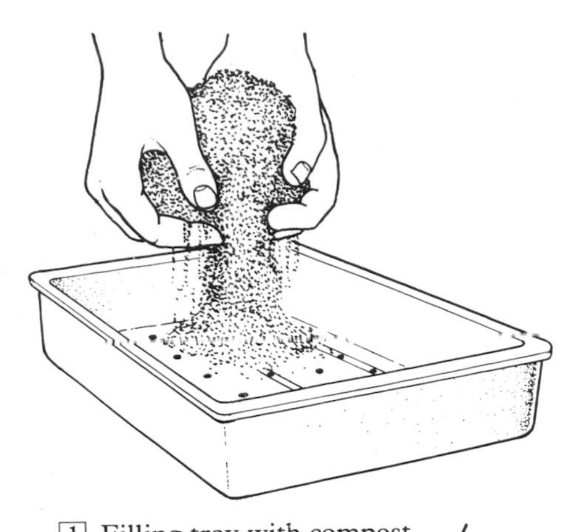

[1] Filling tray with compost.

[2] Firming compost with presser.

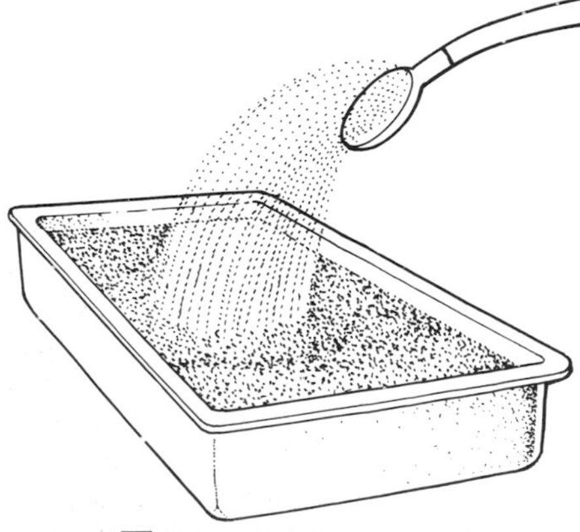

[3] Watering the compost.

Blueprint for seed sowing

A seed is a brand new, baby plant in a dormant state, complete with its own food supply to keep it going for the first few days of growth. All it needs to turn into a seedling is warmth, moisture and air.

[1] Many seeds are best started off in containers, where you can keep a close watch on their progress. Fill a seed tray with sowing-and-cuttings compost: peat-based types (preferably containing a little sand) are the easiest to handle. Heap the compost well up in the centre of the tray, then push it into the corners and spread the heap out until the whole tray is filled level with its rim. Strike off any excess compost by running a piece of wood along the edges of the tray.

[2] Use a wooden presser to firm the compost lightly so that you end up with an even, level surface just below the top of the tray. Make sure there are no hollows, especially at the corners.

[3] Water the filled tray, using a fine rose on the watering can. The rose should be fitted so that the holes face upwards; the water then fountains out of the can giving a very gentle spray. Make sure the rose fits the spout firmly – if it is at all loose it either falls off (disastrous) or dribbles irritatingly, ruining your level compost surface. Leave the water trays to drain for half an hour or so before sowing.

[4] Tap the seed down to the bottom of the packet before tearing off the top. Pinch the lower edge of the packet in the centre to form a V shape to guide the seeds out, then shake them out carefully and thinly to cover the whole surface of the tray. To get the seeds flowing evenly you can tap the side of the packet with your forefinger, or tap the back of the hand that is holding the packet. (Alternatively, tip all the seeds into the palm of your left hand and then tap the edge of that hand with your right hand to steadily dislodge them.)

It's very important to sow *thinly*. Remember that each seed is (we hope) going to develop into a sturdy seedling; try to visualize the space that seedling will need while you are sowing.

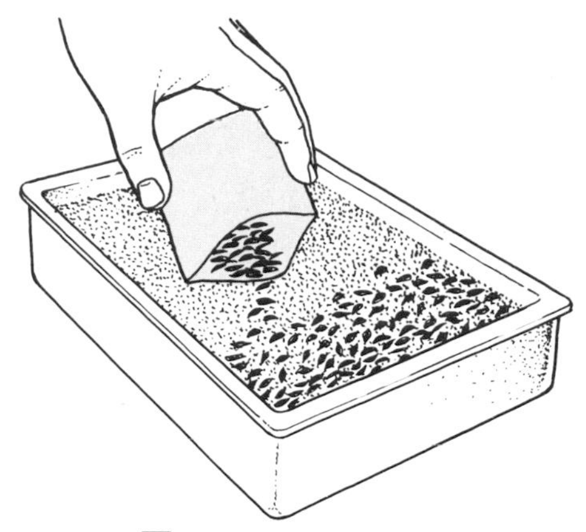

[4] Shaking out seeds.

[5] Most seeds must now be covered with compost to anchor them in place and prevent them from drying out. (Very fine seeds (p. 9) need no compost covering.) Use a sieve to obtain an even covering. With most small seeds it is safest to stop sieving as soon as the last seed disappears from view. A general rule to follow is to cover to the depth of the seed as it is measured from side to side.

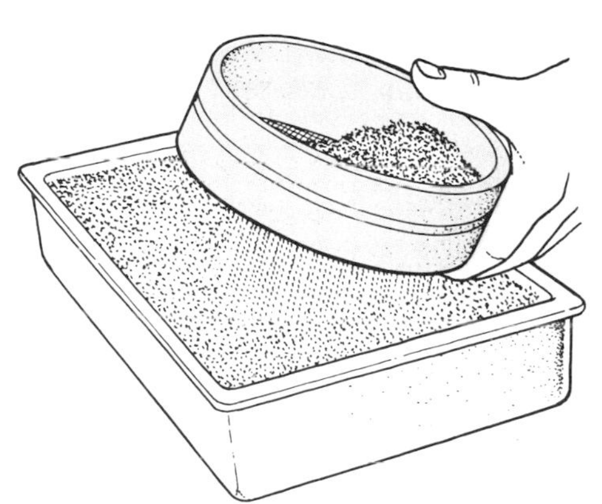

[5] Sieving to evenly cover seeds.

[6] A sheet of clean glass is placed over the sown tray to keep the humidity and temperature high: for extra warmth, a layer of newspaper goes over the top of the glass. Place the covered trays in an evenly warm atmosphere. Different seeds have varying temperature requirements, but 18–21°C (65–70°F) is a good average.

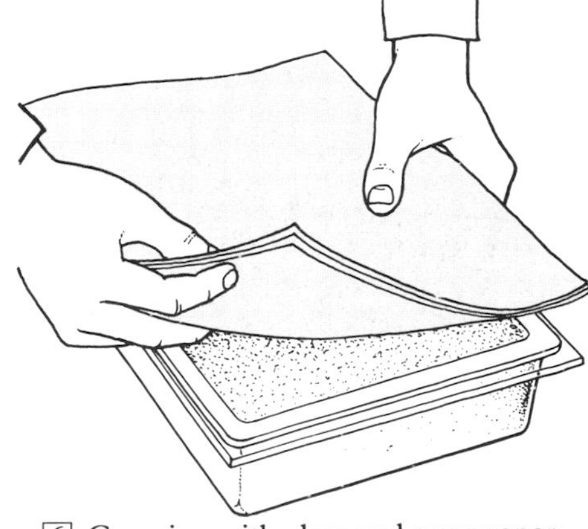

[6] Covering with glass and newspaper.

[7] The time taken from sowing to germination varies enormously. It can be as little as a day, or in some cases more than a year! For most seeds, 14 days is a reasonable period, though the

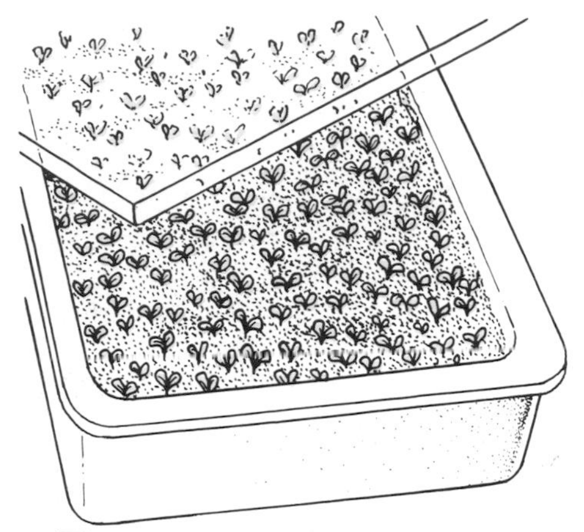

[7] Seeds showing through compost.

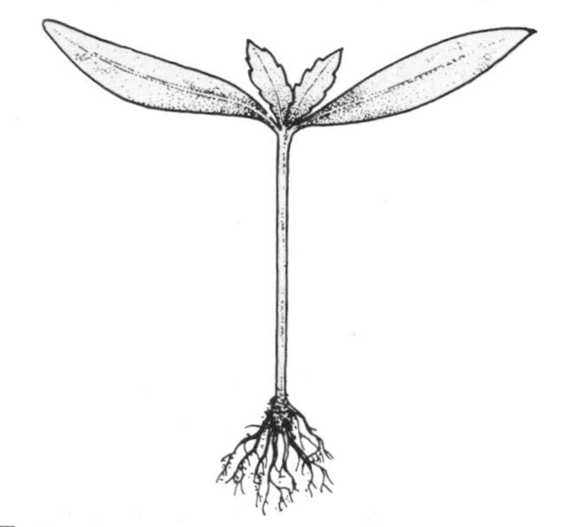

[8] Seedling: seed leaves and true leaves.

[9] Pricking out into tray.

seed packets and catalogues should give you an idea of what to expect. Check the trays every few days, turning the glass over each time to avoid condensation drips. As soon as you see the first few seedlings appearing through the compost, remove the newspaper and move the tray to a light position, but not into direct sunlight.

[8] Remove the glass as soon as the first seedlings touch it, or when the majority of the seeds have germinated. The seedlings will first produce a pair of rather large, coarse seed leaves or cotyledons. The next pair of leaves (called true leaves) have the usually quite different characteristic leaf shape of the young plant.

[9] Once the seedlings are large enough to handle comfortably, they can be moved on to give them more space. Handle them very carefully. Loosen the compost underneath them with a dibber or pencil, and pull the young plants apart by grasping their seed leaves. Never hold seedlings by their stems.

Fill a seed tray with potting compost and dib holes in it at regular intervals (the spacing depends on the seedling size). Still handling the seedlings by their seed leaves, place them in the holes so their roots are in contact with the compost, and very lightly firm them in with the dibber and your fingers. The seed leaves should be almost resting on the compost surface when you have finished.

Water the completed trays with a fine rose and shade the young plants from direct sun for a few days; you can lay a single sheet of newspaper or tissue paper on top of them. Shallow trays of compost dry out quickly in a warm atmosphere, so check them regularly, always watering with the fine rose.

Sowing outside

Some seeds are not sown in containers, but direct into the garden soil – vegetables particularly, but also some flowers. This is either because they germinate so easily that they don't need the extra attention or because they wouldn't like the disturbance involved in transplanting them from their containers.

[1] The soil for sowing should be broken down to a fine, crumbly texture – no large stones or clods. It should also be level, and free of weed growth. Remove all weeds before you start, then on a dry day, rake the surface until it is the correct fine texture, breaking up clods with the back of the rake and removing stones. If the area has only been recently dug or the soil is light and 'fluffy', shuffle all over it to tread it firm. Then using light strokes, rake the surface level. This needs a little practice before you achieve a satisfactory result! Work from the edge of the bed so you don't leave a trough of footprints to spoil the effect.

[1] Raking the seed bed.

[2] Making a shallow drill.

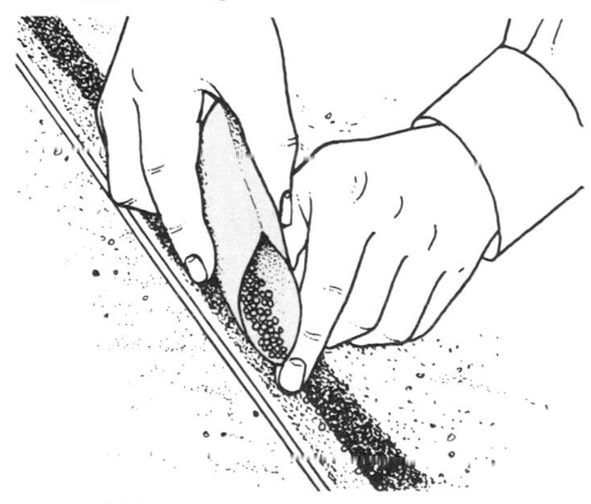

[3] Tapping out seed.

[2] Put down a taut line to mark the row and draw out a shallow drill along it, using the edge of the hoe or the back of the rake. Usually the drill should be ½–1 in (1–3 cm) deep: try to keep the depth as even as you can. If the soil is at all dry, soak the base of the drill before sowing.

[3] Tap the seed out of the packet into the base of the drill, again sowing as thinly as possible. (Alternatively pour all the seeds into the palm of one hand and with the other hand take a 'pinch' of them at a time and scatter them thinly down the row.) Work carefully, as it's often difficult to see the seeds as they fall against the soil. Continue until the whole drill is sown.

[4] Covering seeds by raking.

[5] Firming with back of rake.

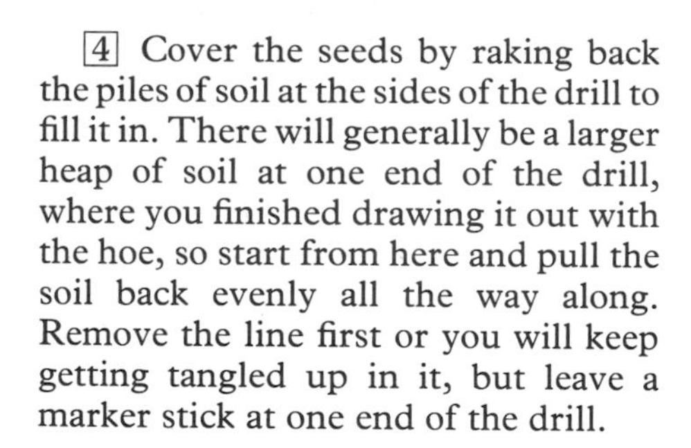

[4] Cover the seeds by raking back the piles of soil at the sides of the drill to fill it in. There will generally be a larger heap of soil at one end of the drill, where you finished drawing it out with the hoe, so start from here and pull the soil back evenly all the way along. Remove the line first or you will keep getting tangled up in it, but leave a marker stick at one end of the drill.

[5] Bring the seeds into good contact with the soil by firming it thoroughly. Use the back of the rake to tamp the soil down.

[6] Label the row clearly with the type of vegetable, variety and date of sowing. You might think you'll remember anyway, but you won't. Use an indelible pencil or the first shower will wash the label clean. (At the end of the season the labels can be cleaned with a damp cloth dipped in scouring powder.) Write all the labels the same way, and put them in the same end of the rows.

Cats can be a problem as they seem to

[6] Label the rows.

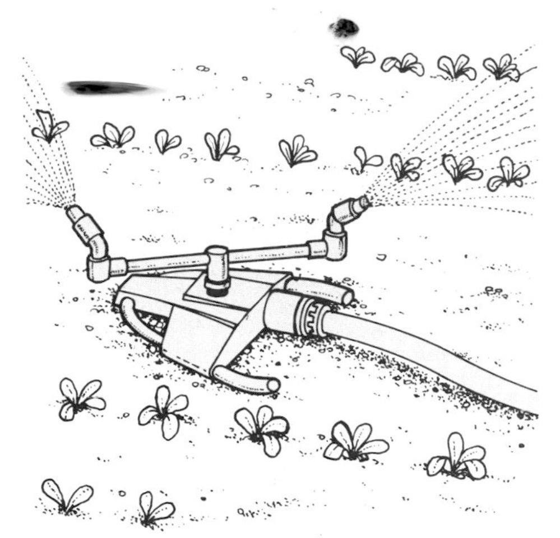

[7] Watering with sprinkler.

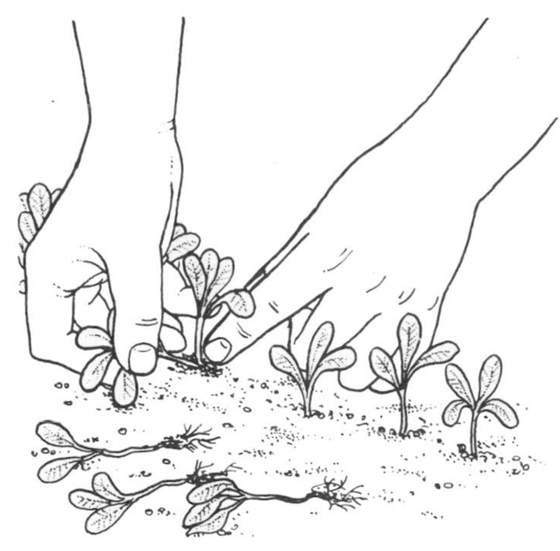

[8] Thinning out seedlings.

find newly prepared seedbeds irresistible; they will dig several large holes before excavating one to their satisfaction. Garden netting, or cotton criss-crossed between sticks just over the soil surface, should protect your newly sown seeds.

[7] Ideally, gentle rain should fall all through the night after sowing, but naturally this hardly ever happens. If there is a dry spell for several days after sowing, the seeds will need watering to stimulate growth.

Watering must be done thoroughly or not at all – there's no point in just damping the soil surface. It takes a lot of heavy watering cans to make any impression on even a small vegetable plot, so wherever possible, use a hose and fine sprinkler. This gives a nice light 'rain', but it should be left in one place for at least half an hour to do any good. An oscillating sprinkler will cover a larger area and should be left running for two hours.

[8] Once the seedlings show through the soil, give them another watering if the dry spell continues (which is unlikely in our climate, but it has been known). Be careful not to damage them by dragging the hose across the rows. Within a short while, the seedlings will be ready to be thinned out to their final spacings (given on the packet).

Thin them when the seedlings are large enough to handle comfortably, but before they are overcrowded, with entangled roots. If you've sown thinly enough you can chop out the unwanted seedlings with a hoe. Use a piece of stick marked off at the correct spacing to guide you until you can judge the distance by eye.

More thickly sown seedlings need careful thinning by hand. Select the strongest, sturdiest seedling at each station, and carefully pull away all the others around it.

Sowing fine and large seed

[1] Seed size varies quite a lot, and some seeds are so small they are literally as fine as dust. Obviously this can pose problems when you are trying to sow them thinly. The way round it is to mix them with a carrier; the best is fine silver sand.

Put a layer of silver sand in the bottom of a clean jam jar. Open the packet of seed very carefully (in a draught-free place!) and tip the contents into the jar. Stir it up with a stick to distribute the seed evenly.

[2] Spread the seed and sand mix evenly over a prepared and moistened seed tray of sowing compost. Do not cover the seed with more compost – just cover the tray with glass and newspaper. Tiny seeds produce tiny seedlings, which will need extra care and attention.

[1] Mixing fine seed with sand.

[2] Spreading seed/sand mix on to compost.

[3] Spacing large seeds in tray.

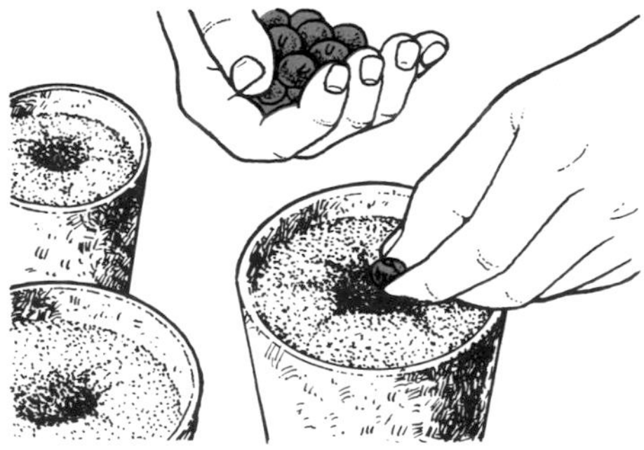

[4] Sowing large seeds in pots.

[3] Large seeds are easier to deal with. They have the advantage that they can be spaced by hand in the seed trays so they won't need pricking out for considerably longer than normal. Remember, the larger the seed, the deeper it should be sown. There's no need to cover it as a separate operation – just push each seed down to the correct depth in the compost.

[4] Some large seeds are best sown individually in pots, as they can then be planted with the minimum of root disturbance, meaning unchecked growth. It's often a good idea to sow two seeds per pot in case of failures. If both germinate, remove one seedling at an early stage.

Chitting and fluid sowing
Often the most difficult part of seed raising outdoors is choosing the right moment to sow. We have no control over the weather, and a spell of poor weather at sowing time can be disastrous. Some seeds are very slow to germinate, and the longer they lie in cold, wet soil, the more chance they have of rotting or being overtaken by weeds. Parsnips, for example, like a long growing season, so they are often sown very early, when soil conditions are likely to be poor; even in good conditions, the seedlings take at least three weeks to struggle up through the soil.

If these difficult seeds can be germinated in a controlled environment before they go outside, their first major hurdle will have been overcome.

[1] The seeds should be given ideal conditions in which to germinate, and a simple sandwich box will provide them. Line the base of the box with several thicknesses of absorbent paper and tip in a cupful of tepid water. When the paper has absorbed as much water as it can, pour off the excess.

Sprinkle the seed thinly over the moist paper and put the lid tightly on the box. Leave it in a warm room.

Within a few days in these warm, moist conditions, the seed will have absorbed water and swollen. The seed-coats will split, and a tiny rootlet will be visible.

[2] As soon as this happens, the germinated seeds must be 'sown' outside without delay. They are very vulnerable at this stage, and the more they are allowed to develop, the less chance you have of transplanting them successfully. Rinse the seed off the paper into a kitchen sieve by holding it under a gently trickling tap.

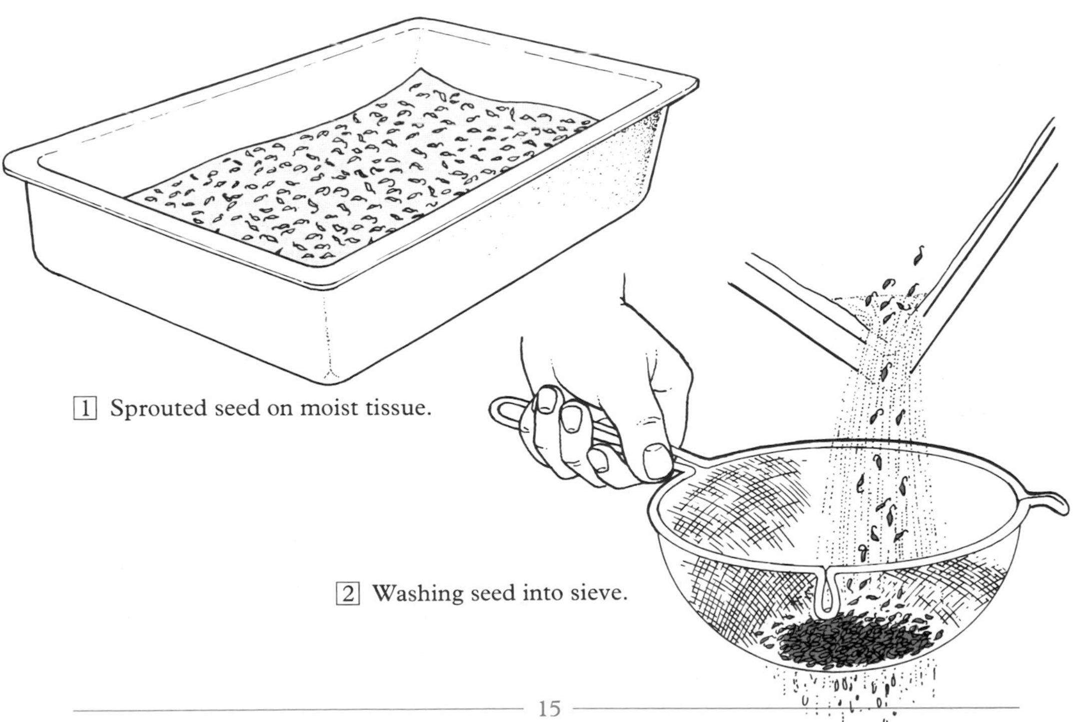

[1] Sprouted seed on moist tissue.

[2] Washing seed into sieve.

3 The germinated seeds are too fragile to be handled normally, and need a carrier that will protect them from drying out and from damage. Wallpaper paste does the job very well. Choose a brand without fungicide (which might damage the seedlings) and mix it up as directed on the packet. Have the paste ready before you wash the seeds out of their box.

Very carefully stir the germinated seeds into the wallpaper paste so they are well distributed.

4 Draw out a drill in the garden in the normal way, but rather deeper, and water it thoroughly. Tip the wallpaper paste mix into a strong polythene bag and tie the top. Take the bag out to the drill and snip one corner off with a pair of scissors.

The paste and seeds can now be gently squeezed out of the corner of the bag as if you were icing a cake, drawing the bag along the bottom of the drill. Once you have finished, draw the soil back over the drill in the normal way, firming it very lightly.

It's very important that the seed and paste are not allowed to dry out, so keep the drill well watered. It shouldn't take long before the seedlings show above the soil, and they can then be treated like any normally sown crop.

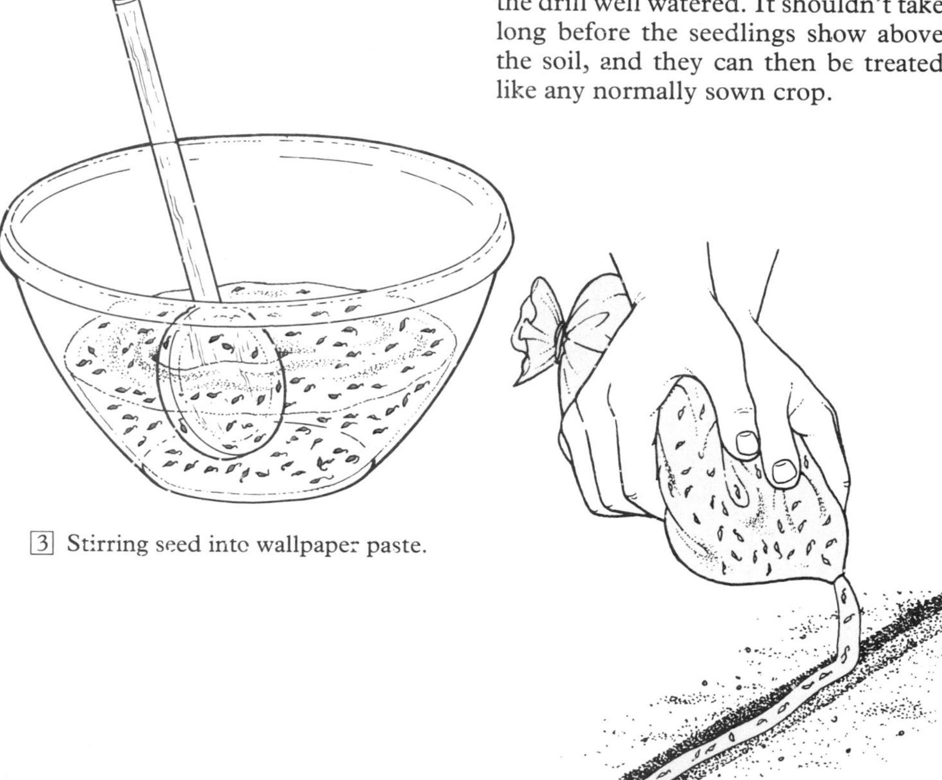

3 Stirring seed into wallpaper paste.

4 Squeezing seed/paste along drill.

Stratification
In natural conditions, most seeds ripen and fall to the ground in autumn. If they germinated straight away, the young seedlings would have to face very tough winter conditions almost immediately, and few would survive. Nature uses various cunning methods for preventing the seeds from germinating straight away – infuriating when we take the seeds out of their natural conditions and try to germinate them ourselves! As long as we know the set of natural triggers for germination we can provide them ourselves and all will be well.

[1] Most seeds in berries and fruits have tough seed coats that need stratifying by cold. Gather the berries or fruits in autumn, when they are fully ripe. Use a block of wood on a wooden bench to crush the berries, breaking down the pulp and exposing the seed coat.

[2] The crushed fruits can now be mixed with approximately twice their bulk of silver sand, stirring the mass thoroughly together. Put the mixture in large pots or other containers with some form of drainage.

[3] An alternative method is not to crush the berries, but to put a layer of sand in the base of the pot (cover the drainage hole with a piece of perforated zinc or plastic netting), then a layer of whole berries, then a layer of sand and so on until the pot is full. Finish off with a layer of sand. It takes a little longer for the frost to get to work on the seedcoats with this method, but for soft-skinned fruits in particular it usually gives equally good results.

[1] Crushing berries.

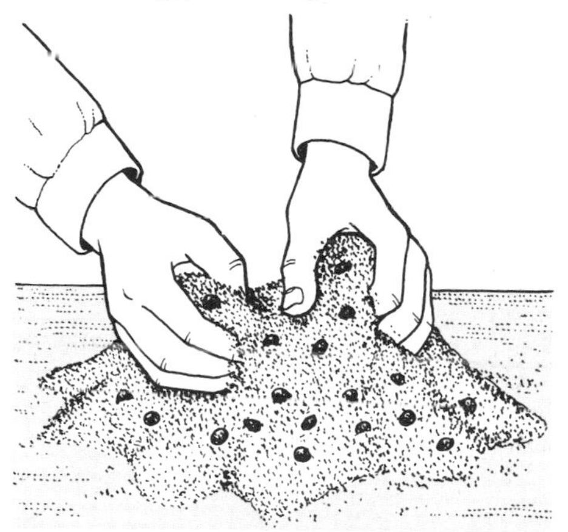
[2] Mixing crushed berries with sand.

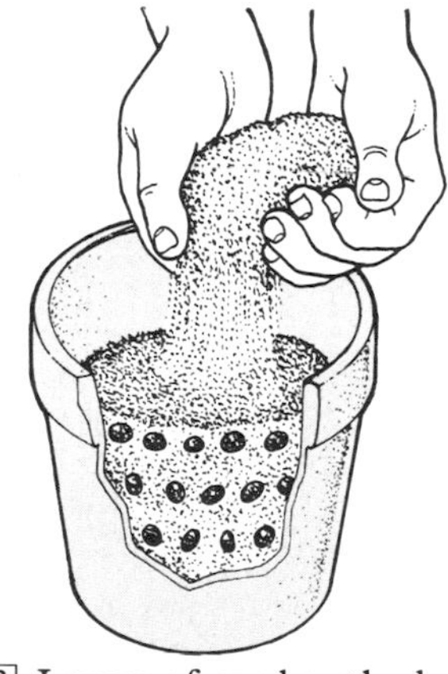
[3] Layers-of-sand method.

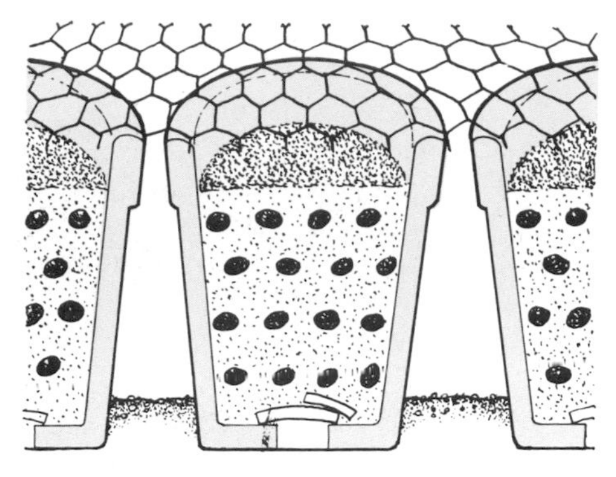

[4] Wire protection against mice.

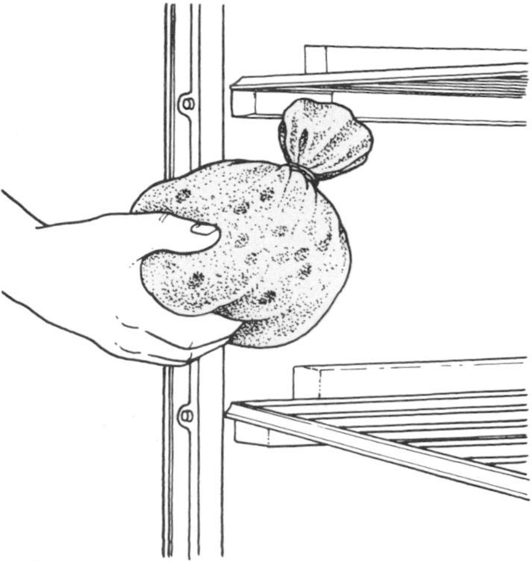

❀ Chilling seed/compost mix in fridge.

☼ Overwintering seed in garden frame.

[4] The pots must now be left outside in an exposed position for the cold winter weather to do its job. Set the bases of the pots in sand or soil for stability, and cover the tops with wire netting or mice will think they've come across a treasure trove. The only other protection needed is against excessive rain (don't set the pots under the drip line of trees or shed roofs, for example). The alternate freezing and thawing will break down the seedcoats and prepare the seed for germination.

Different seeds need stratifying for different lengths of time, but once the correct time has elapsed, the seed and sand mix can be sown in drills in a special nursery seedbed (usually the following spring).

❀ Some seeds don't need prolonged cold treatment, just a period of chilling. Mix the seed with moist, sandy sowing compost in a plastic bag, and tie the top. Put the whole bag into a domestic refrigerator (not the freezer compartment) for the required length of time.

☼ Several types of seed can be sown as normal, in an ordinary seed tray, in autumn. The tray is then left outside through the winter; an open garden frame is a good position. Protect the compost from excessive rain by putting another upturned seed tray over the top, weighting it down with a stone. Check that the compost does not dry out. The seed should germinate the following spring.

Seed coats can stop a seed germinating by preventing it from taking up water. Sometimes it is only necessary to scratch and rupture the seed coat to overcome this.

The scar on a seed (called the hilum) is the point where the root and shoot will emerge, so be careful not to damage this area. Hold the seed firmly, and on the opposite side to the hilum, nick the seedcoat lightly with a sharp knife.

Some seed coats actually contain chemical substances which prevent the seeds from germinating. These 'inhibitors' are, in nature, gradually washed away by rain. Soaking the seed in plain water for a day can dissolve out inhibitors: it also softens any tough seed coats and helps germination in that way.

Seeds may need more than just an overnight soak to soften seedcoats and dissolve inhibitors, and one method involves using very hot water. Put the seeds into a cup and mark off their volume, then tip them into a heatproof basin. Add two to three times their volume of water from a kettle that has just boiled. Don't use any more water than this or you might harm the seeds. Leave them to soak in the water for a day.

Another way to get rid of particularly stubborn inhibitors is to soak the seeds for a day, then tip them into a kitchen sieve and wash them under a gently running tap for half an hour. Cyclamen benefit from this treatment.

Many of these processes won't make the difference between success and total failure, but should certainly improve the percentage and speed of germination.

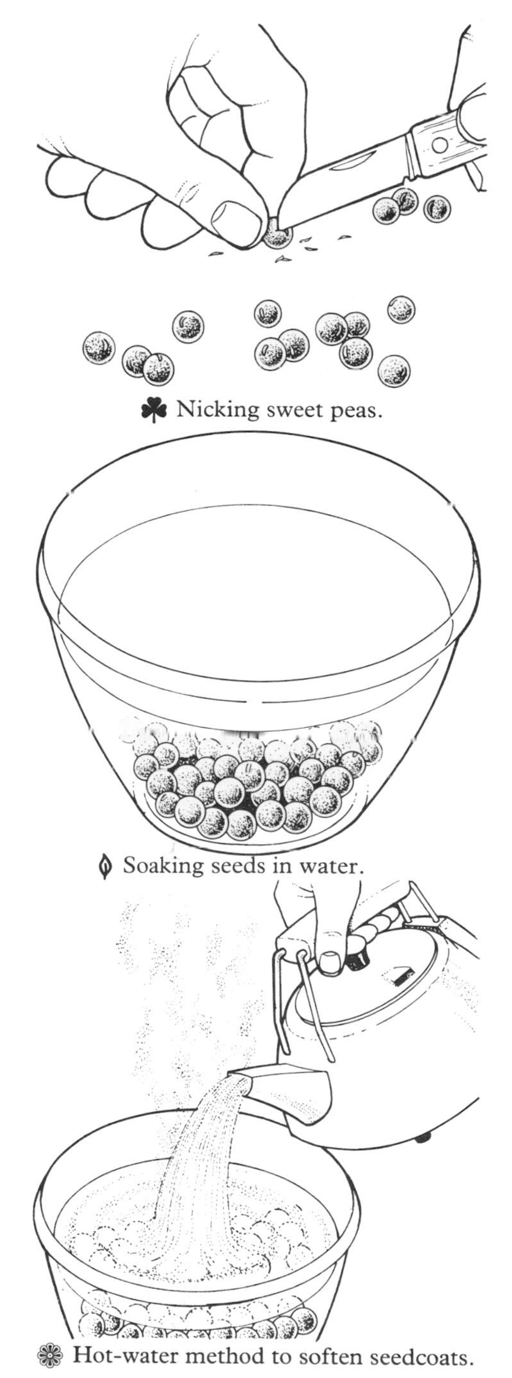

Nicking sweet peas.

Soaking seeds in water.

Hot-water method to soften seedcoats.

❀ Adding silica pack to bean seeds' jar.

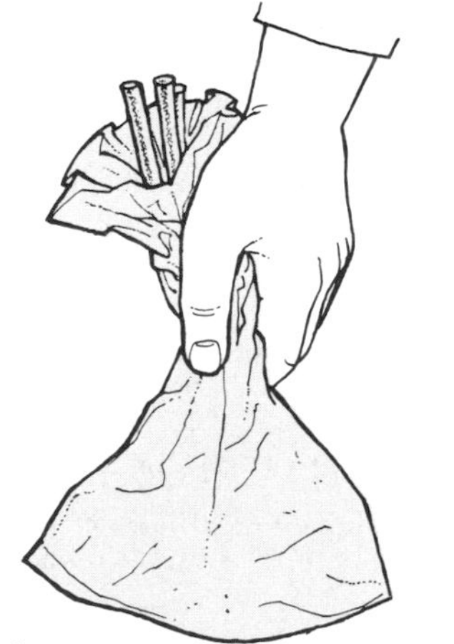

☼ Using paper bag to catch seeds.

♣ Collecting berries from tree.

Saving your own seed

Most of our garden plants produce seeds, and its always fun to grow something from seed you have saved yourself. Remember, though, that cross pollination is likely to give some surprises with the resulting plants.

❀ Runner beans have flowers that tend to pollinate themselves, so the seed they produce will be the same variety as the parent plant. Because bean seed is quite expensive and easy to collect, it's worth saving your own.

Mark one or two bean plants that are producing a particularly good crop earlier in the season, and stop picking beans from these plants. Let the remaining pods ripen fully, until the skins have turned papery and changed colour, while the beans inside are quite hard and dry. Shell out the ripe beans in October and put them in a clean, dry, screw-topped jar. Add a packet of silica gel to absorb any moisture, screw the lid on tightly, and label the jar. Keep it in a cool, dark place until next spring.

☼ The seeds produced in flower seed heads are often automatically scattered as soon as they are ripe. Tie a paper bag over the maturing seed heads to collect the seed: cut the stems and hang them upside down for the seeds to fall into the paper bag.

♣ Collect berries and other fruits from trees and shrubs when they are fully ripe. (You may need to protect them from birds.) Most of these will need stratifying as already described.

What might go wrong

Seed fails to germinate

In containers •Double check expected germination time with seed suppliers. •Check temperature – too high may kill the embryo, too low will delay emergence of seedling and may cause rotting. •Has the seed taken up water? It should be plump and swollen. Seeds may need special pre-sowing treatment (see stratification, p. 17). •Check age of seed. •Make sure compost is not waterlogged or dust dry.

Outside •Cold wet conditions cause rotting of seed. In poor weather, especially early in the season, treat seeds with a fungicidal seed dressing before sowing. •Mice often eat seeds, particularly peas and beans. Deter them by covering the sown drill with prickly leaves and stems before filling in with soil. •A dry spell after sowing will delay emergence. Water if necessary.

Seed germinates but seedlings are unsatisfactory.

Over-thick sowing gives crowded seedlings which become tall, pale and spindly. •Damping off, a fungal disease, causes seedling stems to collapse at soil level; often caused by thick sowing. Other fungal diseases may show as fluffy mould growth. Water remaining seedlings with fungicide solution, such as Cheshunt compound and decrease humidity. •Patchy germination may be caused by uneven, lumpy soil surface at sowing time. •Stretched, pale, sickly looking seedlings need better light conditions. •Brown or yellow scorch marks on leaves may be caused by bright, direct sunlight. •Seedlings that shrivel up and die at an early stage may have been sown too deeply. The same symptoms may be caused by erratic temperatures, or the compost drying out shortly after germination.

Plants which can be increased from seed

Very nearly all plants produce seed – it's a natural form of reproduction. Any seed catalogue will show you the incredible range available. Some plants are not quite as suitable for propagation from seed as others, however, for various reasons: •Some seeds are slow and difficult to germinate. •Seed may be sparsely produced (therefore expensive and difficult to obtain). •Seedlings often show variation – sometimes quite considerable – from the parent plant. •Some seedlings can take many years to reach an acceptable size for the garden. •Seed-raised plants may have a different habit of growth and foliage from mature plants for some years – known as having a 'juvenile form'. •Some plants, apples and pears, for example, have such a mixed genetic background that plants grown from their pips will give totally new varieties. •Seedlings may be so vigorous it will take years before fruit trees settle down to regular cropping.

But seed is still a reliable and cheap method of producing large numbers of plants, particularly garden flowers and vegetables.

Seeds requiring stratification

Seeds of most trees and shrubs can conveniently be stratified from the autumn, when they are collected, to the following spring. Some which need a longer period are: *Acer campestre* (field maple) 18 months; *Crataegus* (hawthorn) 16 months; *Ilex* (holly) 16 months; *Carpinus* (hornbeam) 12 months; *Tilia* (lime) 18 months; *Taxus* (yew) 16 months; and *Fraxinus* (ash) 18 months.

Seeds requiring less than two months' stratification may alternatively be pre-chilled in a refrigerator. They include *Abies, Betula, Cedrus, Fagus, Malus, Morus, Nothofagus, Prunus, Pyrus, Rhododendron, Rhus*.

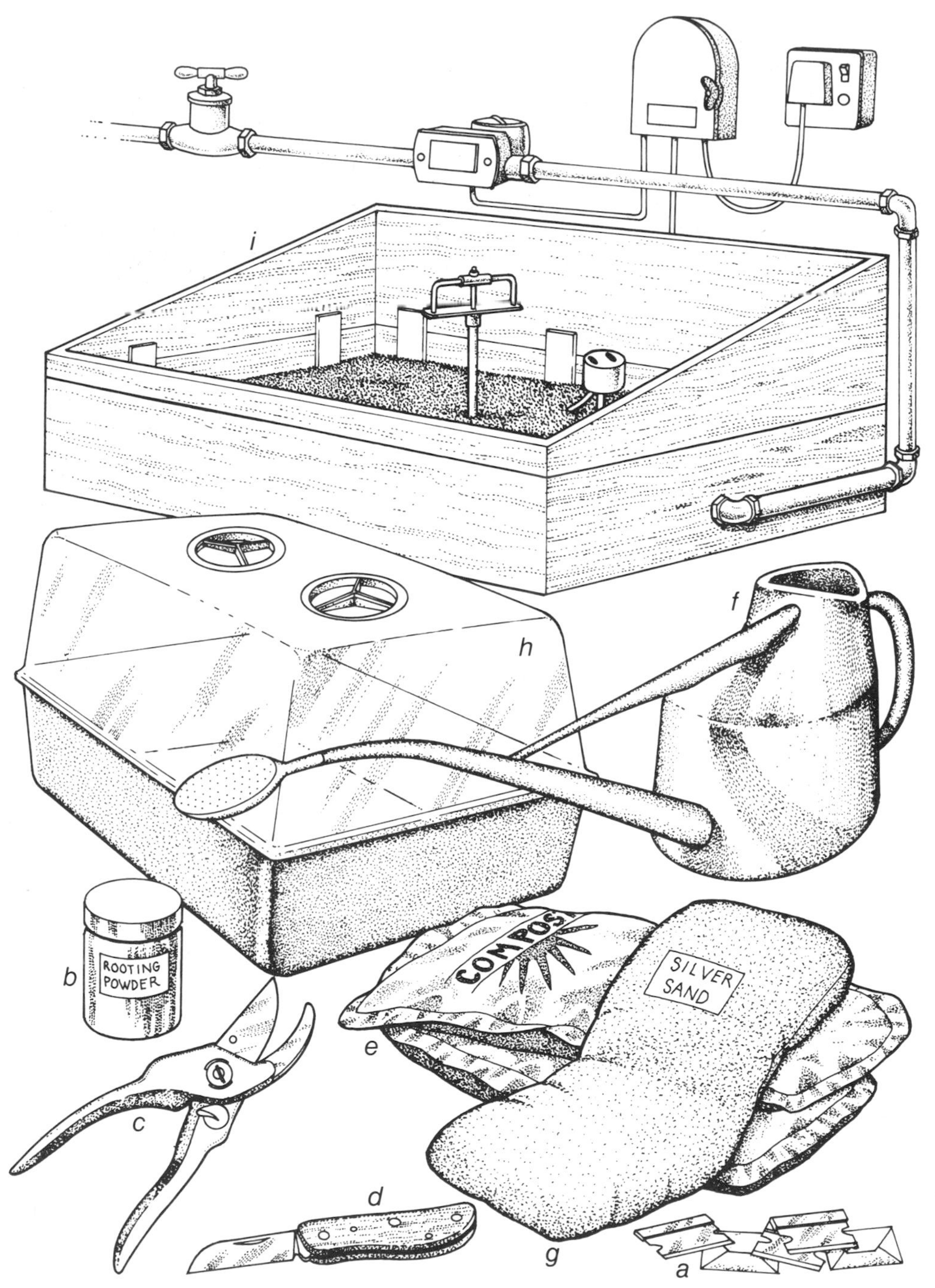

Equipment for cuttings: for details see text opposite.

2
CUTTINGS

Equipment

Much of the equipment you need for seed raising will also come in useful for taking cuttings, but there are a few special items, too.

Disease (in the form of rots) is one of the commonest causes of failure with cuttings. It generally starts in the bruised, crushed tissues at the base of the cutting, and making a clean cut helps to prevent it. For soft cuttings, razor blades (*a*) are ideal. They can be fitted into special plastic covers for safer handling, or one edge can be covered with a piece of masking tape.

Rooting powders (*b*) contain hormones that the plant uses in the formation of roots, and most brands also contain a fungicide – almost more important than the hormones, as it helps prevent those deadly rots. Many plants root happily without rooting powder, but it's never a bad idea to use all possible aids to success.

A good pair of secateurs (*c*) is needed for taking hardwood cuttings, and a well sharpened, straight-bladed knife (*d*) will come in useful, too.

The same types of compost (*e*) and pots you use for seed sowing will be fine for cuttings, too. Again, little nutrient is needed in the compost, and trays and half pots will hold the shallow depth of compost needed. A fine rose on the watering can (*f*) will settle the cuttings in to the compost without washing them away.

Good drainage is essential, especially round the base of the cutting. A layer of silver sand (*g*) on top of the compost will slide down into the hole when the cutting is inserted, so improving drainage around the stem.

Leafy cuttings lose water rapidly through their leaf surfaces, and having no roots, find it difficult to take up enough water from the compost to replace it. This is why they soon flag, especially in hot sunshine. Any way of keeping the air round the top of the cuttings fully charged with moisture – in other words, providing a humid atmosphere – will make the cuttings a lot happier, and let them get on with the business of rooting.

A simple way of providing a humid atmosphere is to enclose the whole pot or tray of cuttings in a plastic bag. Use a couple of canes to support the bag and stop it lying on top of the cuttings. Although this is very effective at keeping a moist atmosphere, lack of fresh air makes fungus diseases like mildew and botrytis likely. A rigid plastic propagator top with adjustable ventilators (*h*) is a much better method – the tops are cheap and long lasting.

The ultimate device for rooting cuttings is a mist unit (*i*). This has soil warming cables to provide the bottom heat that so many plants like, and a fine spray of water is delivered over the tops of the cuttings at regular intervals by a jet in the centre of the unit. An electronic sensor placed some distance from the jet senses how dry the atmosphere is, and turns the water on accordingly.

[1] Select a strong, healthy shoot.

[2] Adding silver sand to compost.

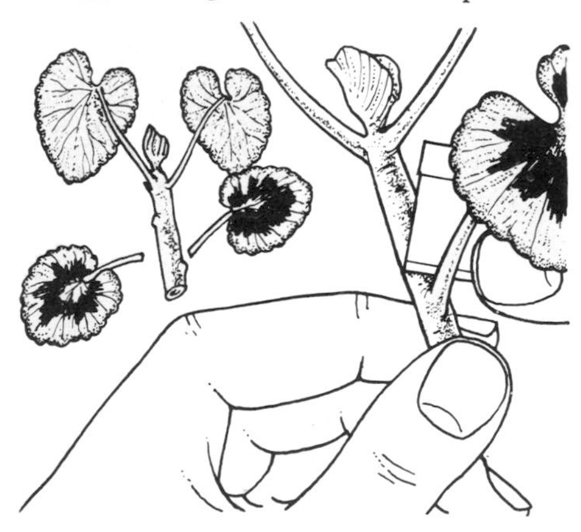

[3] Trimming cutting.

Blueprint for (softwood) cuttings
Cuttings are portions of plants which are encouraged to develop roots or shoots (or both), and become complete plants on their own. Unlike seedlings, plants propagated by cuttings will be identical to their parents. Softwood cuttings are simple to take and give quick results.

[1] Choose a strong, healthy shoot of the current years growth, pest free and preferably without flower buds. All shoots are best cut from the parent plant with a sharp knife to avoid damage. Cut the shoot longer than you need and make the cut just above a node (where the leaf joins the stem). The stock plant will then make new growth from the buds in the leaf axils.

[2] Fill the tray or pot with seed and cuttings compost. Firm the surface lightly with a flat wooden presser, so that it is level and just below the rim. Sprinkle a layer of fine silver sand evenly over the compost surface. This will improve the drainage immediately round each cutting.

[3] Use a razor blade to trim the cutting immediately below a node, so that the entire length is about 2–3 in (5–8 cm). Remove the lower leaves and, if there are any, the stipules (small papery projections on the stem at the base of each leaf stalk). If these are left on they will be below soil level once the cutting is inserted, and are likely to rot. Pinch out any flower buds.

4 Tip a little rooting powder into the lid of the container and dip the base of the cutting in it. Tap the stem lightly on the rim to knock off any excess powder: a light dusting is all that should remain.

Rooting powder contains artificial 'hormones' which help to speed up the rooting process. Choose a brand of rooting powder which contains a fungicide to prevent disease from attacking the vulnerable base of the cutting.

5 Make a shallow hole in the compost with a pencil or small dibber and insert the cutting, making sure the base is in contact with the compost. It's all too easy to end up with an air space underneath, which means all your efforts will result in complete failure! Firm-stemmed cuttings can be pushed directly into the compost without dibbing a hole first. Firm the cutting in gently but thoroughly with your fingers.

Fill the tray or pot with cuttings, spaced out evenly so the leaves don't touch each other, and water them well with a fine rose. Cover with a plastic propagator top and leave the tray or pot in a warm, lightly shaded place.

6 Pelargoniums root quickly, and within 14–21 days you should see the tips of the cuttings take on a fresh green 'growy' appearance. A *very gentle* tug will tell you whether or not the cuttings have rooted – if you feel resistance, roots are there. Once there is a good ball of roots, pot the plantlets up individually in 3½ in (9 cm) pots of peat or loam-based potting compost.

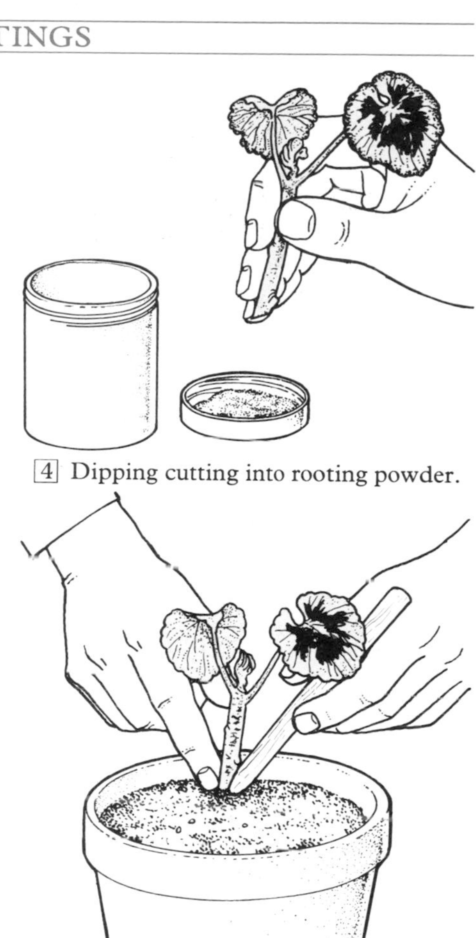

4 Dipping cutting into rooting powder.

5 Firming in cutting.

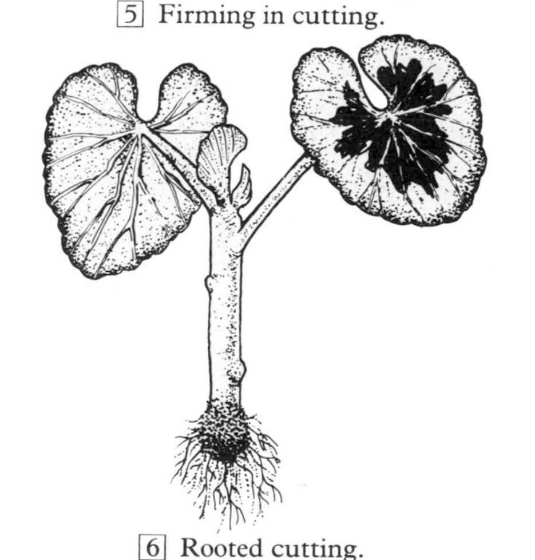

6 Rooted cutting.

[1] Tearing off shoot with heel.

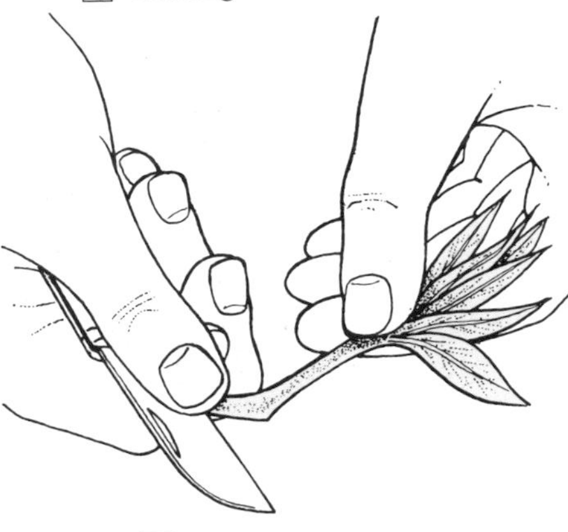

[2] Trimming shoot.

[3] Dipping shoot into rooting powder.

Semi-ripe cuttings
Semi-ripe cuttings are taken in summer, when stems have just begun to harden and ripen. They are rather more resilient than softwoods, rooting fairly easily but not quite so quickly. Many shrubs are propagated by this method.

[1] Choose a side shoot of the current season's growth which is firm at its base, but still pliable. If the shoot is less than 6 in (15 cm) long, tear it away from the main stem with a small heel of rind; otherwise cut it with a pair of secateurs, making it 3–6 in (8–15 cm) in length.

[2] Trim the base of the cutting with a sharp knife, neatening the heel and removing lower leaves and leaf stalks. Because a cutting loses a lot of water through its leaf surfaces it's a mistake to leave too much foliage; usually two pairs of fully expanded leaves is about right. Large-leaved plants can be reduced to one pair plus the terminal bud. Leafy cuttings wilt quickly once they have been cut, and may not be able to take up sufficient water to recover, which rather puts paid to their chances of rooting.

[3] Dip the base of the cutting into hormone rooting powder and tap off the excess. Some brands of rooting powder are available in different strengths for different types of cutting – soft, semi-ripe or hardwood – others are general purpose. Whichever type you use, only leave a thin dusting on the stem. Clog the base with too much powder and you'll just hinder the rooting process.

[4] Prepare a tray of rooting compost covered with sand as in the blueprint, and insert the cuttings in the same way. Firm, and water them in thoroughly with a fine rose.

[5] A garden frame with a base of sand is a good place to root semi-ripe cuttings. Paint the glass with greenhouse shading material or cover the frame lights with green shade netting to prevent the cuttings from being scorched by bright sunlight. Keep the frame closed for the first few days to provide high humidity, but ventilate if the temperature climbs above 24–27°C (75–80°F). Water with a fine spray as soon as the compost begins to show signs of drying. Watering with a fungicide solution once or twice will help to stop diseases thriving in the moist, warm conditions.

[6] Many semi-ripe cuttings will be well rooted by the autumn, but then growth stops for the dormant season. Remove fallen leaves, but unless the cuttings are very advanced, they can be left in their trays in the frames over winter. In very cold weather, cover the frames with sacking or hessian matting.

Growth will start in earnest in spring, and the rooted cuttings will soon show signs of life. Ventilate the frames as the weather becomes milder to harden the cuttings off, then plant them in 'nursery rows' in a sheltered spot in the garden. During their first growing season, keep them weeded and well watered so they have made strong plants by the following autumn. They can then be lifted and transplanted to their permanent positions.

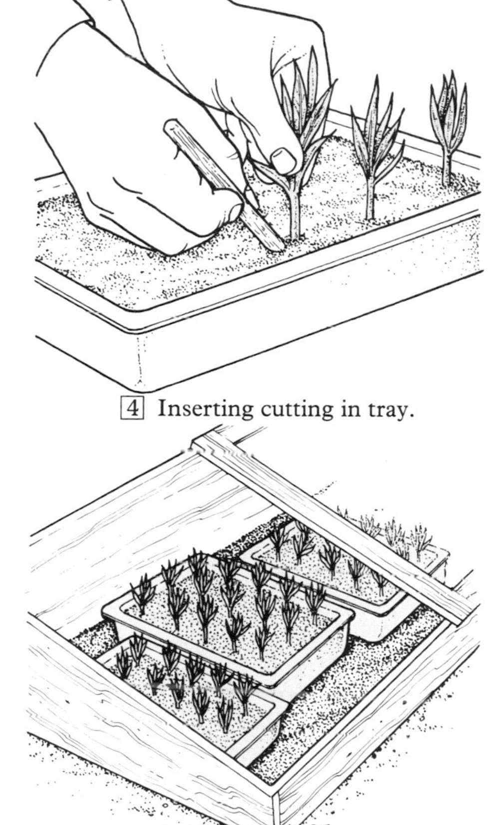

[4] Inserting cutting in tray.

[5] Putting tray in cold frame.

[6] Rooted cutting.

Hardwood cuttings

Hardwood cuttings are made from fully ripened wood, taken during the dormant season. Because the stems are dormant, they don't need special care like leafy cuttings do, so they are very easy. They do require patience, because you won't know what success you've had until the following spring.

[1] Taking suitable shoot.

[1] Take a suitable healthy shoot of the current year's growth from the parent plant early in the winter, cutting out the entire shoot. Don't leave any snags to die back on the parent plant: where possible, cut just above a bud. On deciduous plants all the leaves should have fallen; any dead leaves still clinging to the stems should be brushed off.

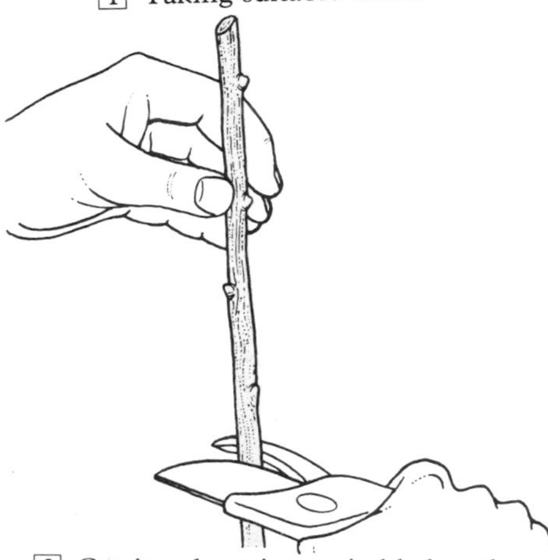

[2] Cutting shoot into suitable lengths.

[2] One branch of many trees and shrubs will give several cuttings, though the piece of branch nearest the tip often roots less successfully than that at the base.

Cut the stem into lengths of 6–10 in (15–25 cm), making a straight cut at the bottom of each length and a sloping cut at the top, if it will help you remember which end is which. Discard any thin, soft tips – they are usually not worth bothering with. With hardwood cuttings you can make the cuts anywhere on the stem; they don't need to be under a node.

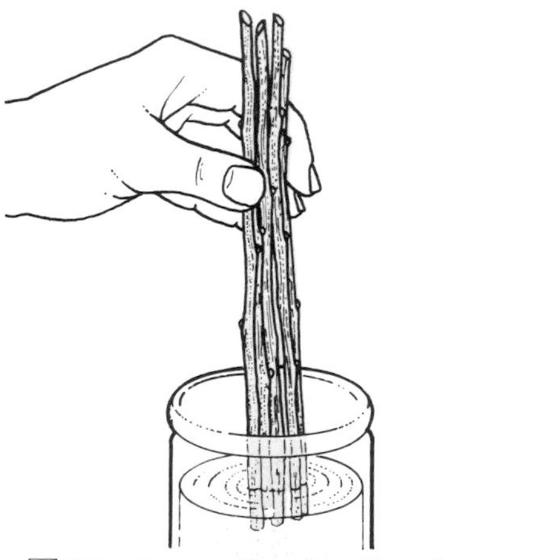

[3] Dipping cuttings into liquid rooting hormone.

[3] Treat the bases of the cuttings with rooting hormone – for hardwoods this is often formulated as a liquid. Make a small bundle of cuttings with their bases level and dip them into the prepared liquid for about five seconds, or as directed on the pack.

[4] Hardwood cuttings are rooted in the open ground – they don't need mollycoddling in frames and propagators. The ground should, however, have been properly prepared, being dug over thoroughly in autumn, with some peat and sand incorporated. Perennial weed roots should have been removed.

Make a slit trench in the prepared and firmed ground by inserting the space vertically and pushing it forwards slightly. Using a line to guide you will ensure neat, straight rows.

Insert the prepared cuttings about 4–6 in (10–15 cm) apart. They should be inserted deeply, so that only the top two or three buds are left showing above the ground. Make sure the cuttings are put in the right way up – it's usually obvious, but on some plants you could make a mistake if you stop concentrating on what you're doing – upside down cuttings don't root very well.

[5] Firm the cuttings into the trench by pushing the soil back with the ball of your foot. Label the row, then leave the cuttings to form callus tissue – a corky white substance that 'heals over' the base – through the winter. During winter heavy frosts may disturb the soil and 'lift' the cuttings; in this case, refirm them gently by treading the soil back.

[6] During the following spring, roots will grow from the callus tissue, and the buds above the soil will start to expand. Don't disturb the cuttings, but keep them watered and free of weeds through the following growing season. By the autumn the young plants should be large enough to lift and transplant to their final positions.

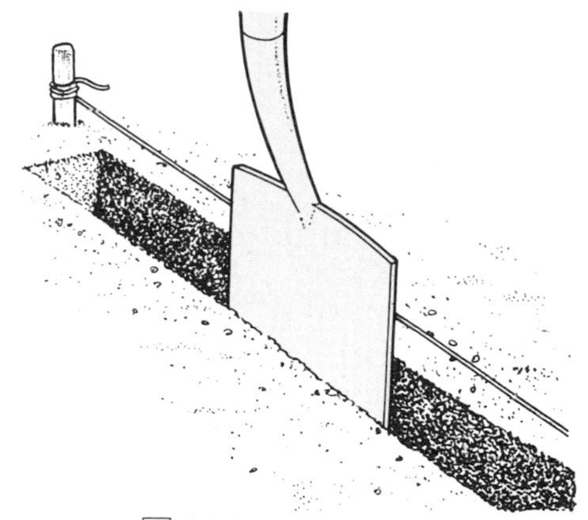

[4] Making slit trench.

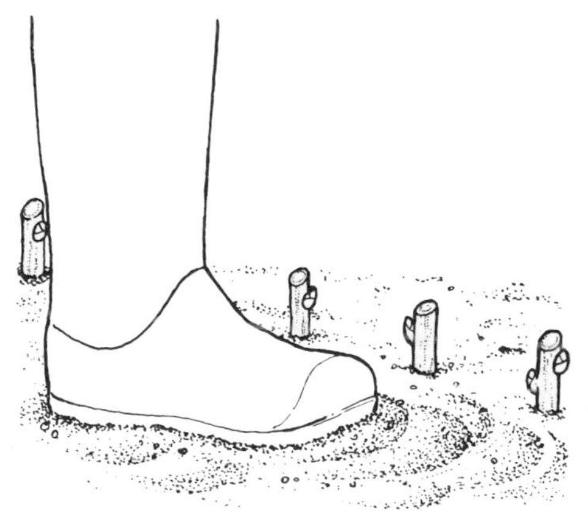

[5] Firming cuttings into trench.

[6] Transplanting rooted cutting.

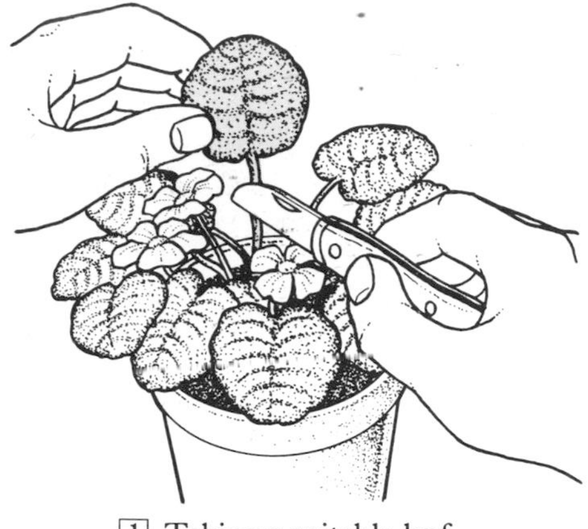

[1] Taking a suitable leaf.

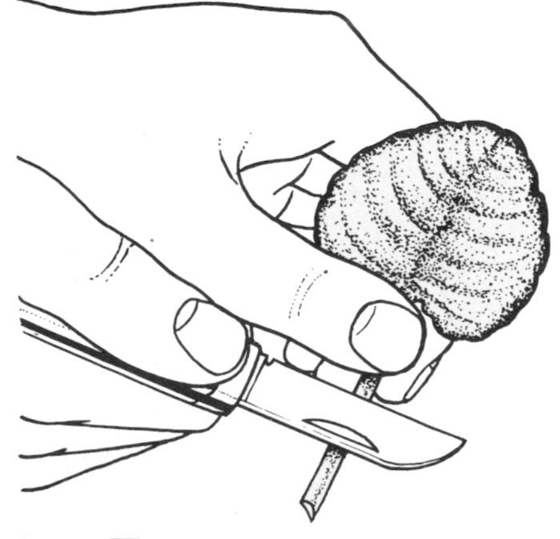

[2] Trimming leaf stalk.

[3] Inserting leaf into compost.

Leaf cuttings

Stems are not the only parts of plants that will produce roots: leaves and leaf stems sometimes will, too, and this is a popular way of increasing many house plants.

Saintpaulias

[1] Saintpaulias, or African violets, will produce several new plantlets from one leaf. A leaf cutting can be taken at virtually any time of year, but spring and summer cuttings root most easily.

Choose a fully open, healthy leaf with no sign of damage or scorch. Cut the leaf off as near the base of the leaf stalk as you can, using a razor blade or sharp knife.

[2] Lay the leaf on a flat surface and trim the leaf stalk so that it is about 1–1½ in (2.5–4 cm) long, with a clean cut at the bottom. Be careful not to bruise the leaf blade as you are doing this, because damaged leaves will rot.

[3] Fill a pot with sowing and cutting compost and firm it down lightly. You can also use an inert material like vermiculite if you prefer – this gives good results. Water the compost thoroughly. Make a hole with a dibber or pencil slightly off centre and insert the saintpaulia leaf so that half to two-thirds of the leaf stalk is buried. Firm gently but thoroughly with your fingers.

Keep the pot in the house, in a warm position out of direct sun. The greenhouse is not the best place for saintpaulia leaf cuttings as they are likely to be scorched by the bright light, but if they can be placed in a shaded part of the greenhouse they should be all right.

[4] Keep the compost just moist, being careful not to splash any drops of water on the leaves. In a short while, roots will form at the base of the leaf stalk, but it will be a few weeks before you see tiny new plantlets pushing up through the soil at the base of the leaf. How long they take to form varies with the season and conditions, but six to eight weeks is normal. Once through the soil they grow rapidly, and when the pot is full of leaves it's time to transplant them.

[5] Nearly all the leaf cuttings will produce a cluster of new plantlets, and these must be separated when they are potted up. Turn the pot upside down and tap the rim on a firm surface to loosen the compost and remove the plants. Once out of the pot, it should be quite easy for you to identify the separate crowns and tease the plantlets apart. Pot each plantlet in its own 3½in (9cm) pot of potting compost, making sure the crown is firm.

The 'mother' leaf can be detached from the cuttings, and if still in good condition, could even be used again; worthwhile if the plant you are propagating is a particularly precious variety. Cut the base of the leaf stalk off with a razor blade and insert it into a fresh pot of compost. If the leaf has started to shrivel and go limp, discard it – before too long you can use a fresh leaf from your new plantlets.

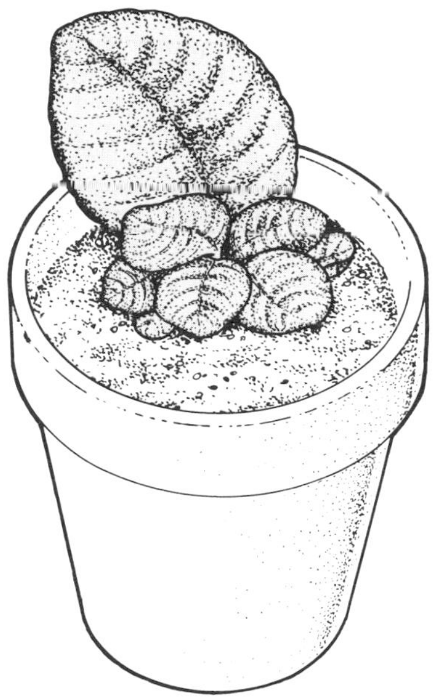

[4] New plants forming.

[5] Separating the plantlets.

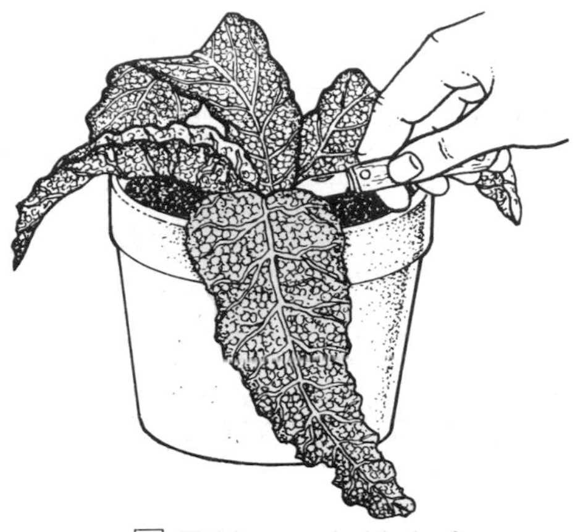

[1] Taking a suitable leaf.

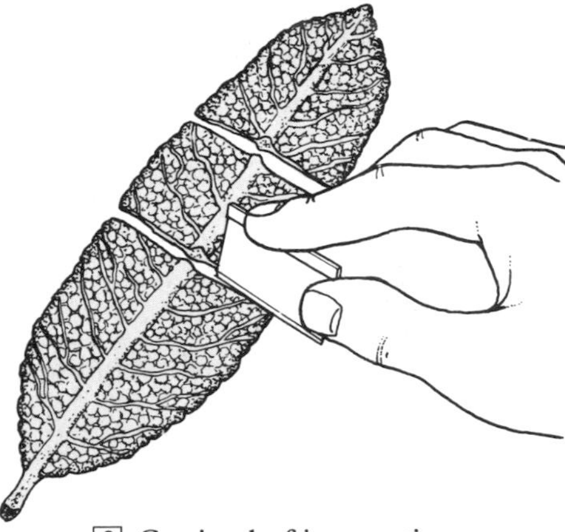

[2] Cutting leaf into sections.

[3] Inserting sections into compost.

Streptocarpus

[1] Streptocarpus plants have long, rather ungainly leaves that are easily damaged, and quickly become brown at the tips and edges. Trim off all brown or yellow leaves late in the winter, and water and feed the plants to encourage strong, healthy growth. In spring, choose a fresh, green but fully expanded leaf for propagation and cut it off at the base. Like saintpaulias, streptocarpus leaf cuttings can be taken whenever there is suitable foliage, but spring and summer cuttings are most successful.

[2] Because the leaves are long, each one can be cut into two or three sections. Use a very sharp knife or razor blade. The lower part of the leaf (nearest the leaf stalk) usually roots more readily than the tip. Brush or dust a very small amount of rooting powder on to the lower cut of each section.

[3] Prepare a tray of seed and cuttings compost and water it thoroughly before inserting the cuttings. Push the leaf sections into the compost for about one-third of their depth.

Keep the tray of cuttings in an evenly warm atmosphere, out of direct sunlight. If the compost starts to look dry, water it by standing the tray in a large bowl of water so that moisture can be drawn up from the base.

The leaf cuttings are often rather slow to root and produce plantlets. Remove any leaves that become limp and brown and are obviously dead.

[4] In due course plantlets should develop at the point where the veins meet the compost surface. Once the plantlets are well developed, remove them from the tray, separate them from the parent leaf and pot them up individually.

[5] Another way of making streptocarpus leaf cuttings is to divide the leaf longitudinally. Again, choose a mature, healthy leaf; it makes things easier if it is a firm leaf which is as straight as possible (streptocarpus leaves have a tendency to curl under). Lay the leaf face down on a smooth surface and on the back you will be able to see the raised lateral veins branching out from the midrib. Each of these veins has the potential to produce a new plant.

Run a razor blade closely up each side of the midrib so the entire midrib is cut out: this is discarded. The lateral veins are now exposed. Dip the cut edges of the leaf in rooting powder and dust off the excess with a soft brush.

[6] Prepare a seed tray of compost as normal and insert the two halves of the leaf, burying their cut edges. Leaves prepared like this are rather more difficult to handle than horizontally cut ones. Wide leaves may need support to keep them upright: a wooden or plastic plant label, snapped in half, makes a useful prop.

Keep the leaf halves in the same conditions as the horizontal leaf sections, and eventually plantlets should form all the way along the leaf where the veins have been exposed. Once they are well developed, tip them out of the tray, cut the mother leaf to separate the plants, and pot them up individually.

[4] Plantlets developing.

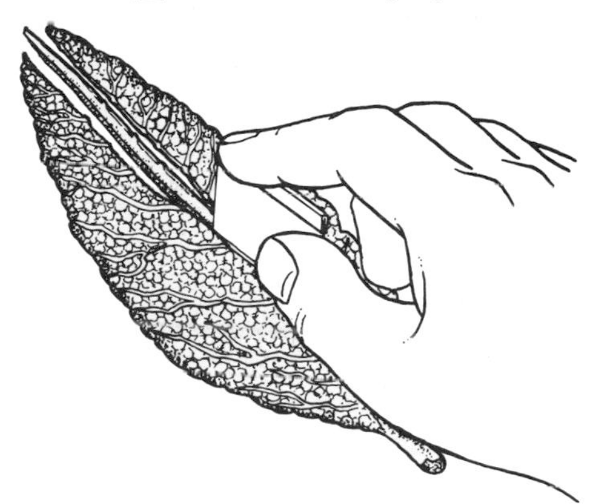

[5] Longitudinal division of leaf.

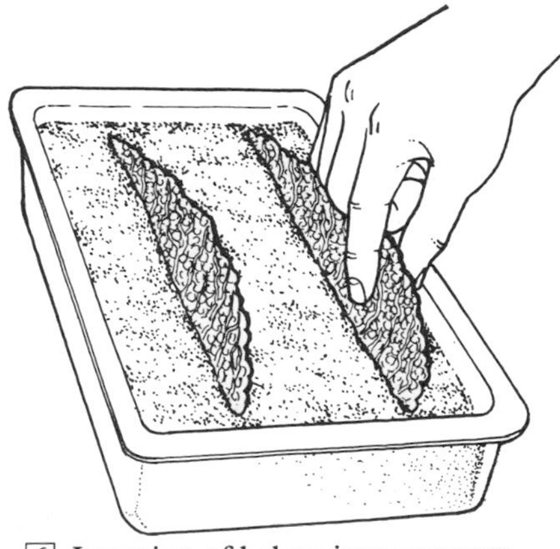

[6] Insertion of halves into compost.

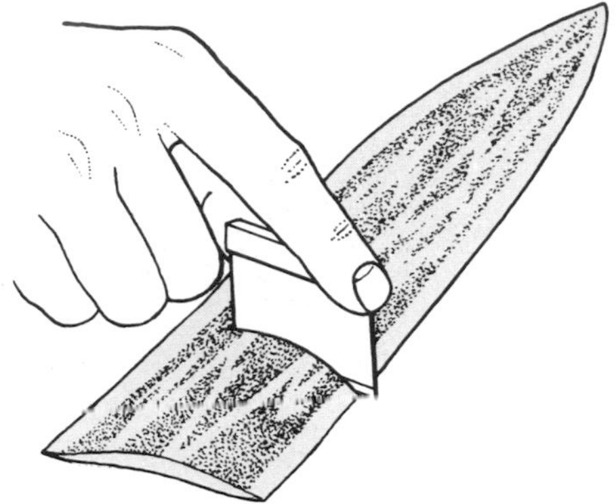

[1] *Sansevieria*: Slicing leaf into sections.

[2] Inserting sections into compost.

Sansevieria

[1] Mother-in-law's tongue is a favourite houseplant with its long, fleshy, spiky leaves. The most popular form is the yellow edged *Sansevieria trifasciata laurentii*: if you propagate this from leaf cuttings the new plants won't be variegated, but will have green leaves marbled with grey.

Cut an entire, healthy, unblemished leaf from the plant and lay it on a flat surface. Cut the leaf into sections 2–3 in (5–8 cm) wide with a sharp knife, and dust the base of each section with rooting powder.

[2] Insert the leaf sections to half their depth in a tray of cuttings compost – be careful to get them the right way up – and firm them in with your fingers. Water the compost, cover the tray with a ventilated propagator top and keep in a temperature of around 18°C (65°F).

Begonia rex

[1] Remove a young but fully developed leaf from a *Begonia rex* and lay it face down on a smooth surface. The main veins stand out prominently on the reverse of the leaf; with a knife or razor blade, make a slit in each vein.

[2] Lay the leaf face upward on a tray or pan of thoroughly moistened cuttings compost and press it lightly into the surface so that the cut veins are in good contact with the compost. It will need holding in position, and ordinary hairpins are quite useful for pegging leaves down. A few pebbles placed on the leaf will also do the trick.

If the compost shows signs of drying, mist it over with plain water using a hand sprayer. Given reasonably warm and humid conditions, plantlets should develop from each cut in the leaf veins.

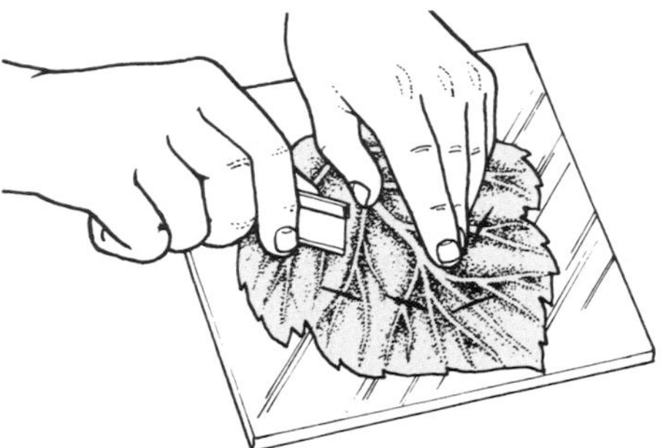

[1] *Begonia*: Cutting slits in leaf.

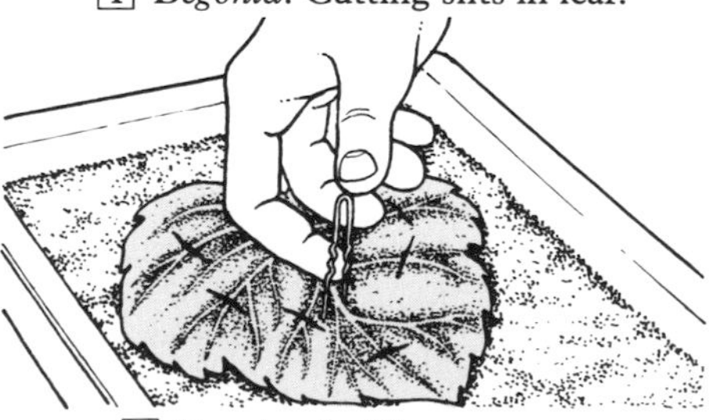

[2] Pegging leaf into compost.

Leaf bud cuttings

Although plants like saintpaulias have the almost miraculous ability to form an entire new plant – leaves, roots and stems – from one small section of tissue, others need a little more of the basic material to work with.

[1] Camellias are tricky plants to propagate, but leaf bud cuttings are worth a try. In August, cut a healthy shoot that has grown that year and is beginning to ripen at the base. This will provide you with material for several cuttings. Each cutting consists of a leaf complete with its axillary bud and a piece of stem about 1 in (2.5 cm) long: make the top cut just above the bud with a pair of sharp secateurs.

[2] Dip the base of the cutting into a hormone rooting powder formulated for semi-ripe cuttings. Tap the cutting gently against the side of the container to knock off any excess powder.

[3] Prepare a tray or pot of cuttings compost with a little fine silver sand added, and firm it down lightly. Insert the cuttings so that the leaf and bud are just above the surface, and firm it into position with your fingers. If the leaf is large, the cutting may need supporting with a piece of split cane. Finally, water the inserted cuttings well.

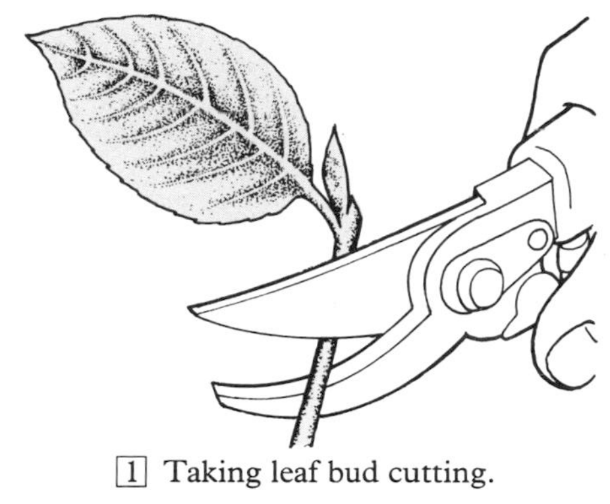

[1] Taking leaf bud cutting.

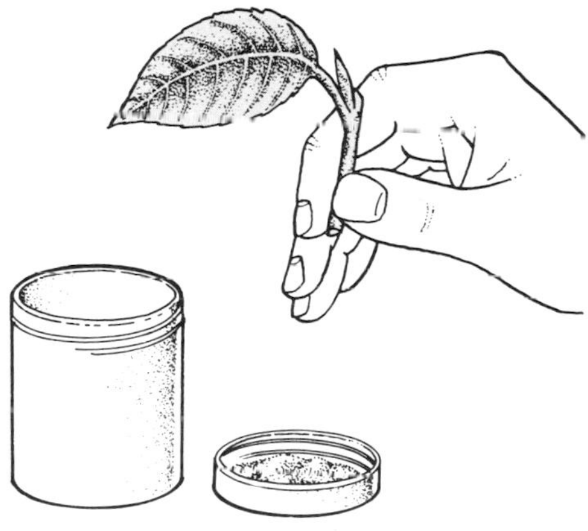

[2] Dipping cutting base in hormone powder.

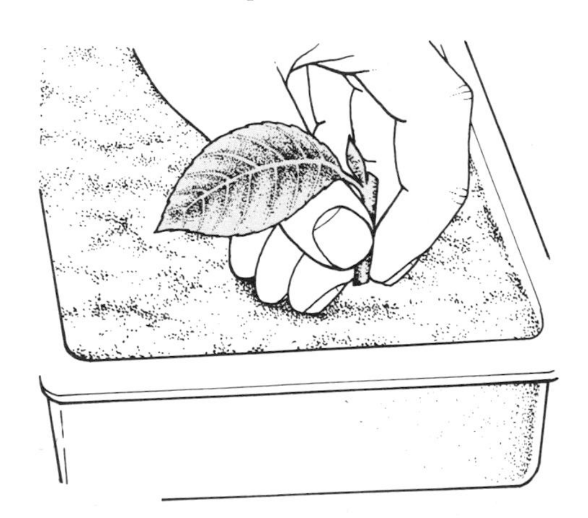

[3] Inserting cutting into compost.

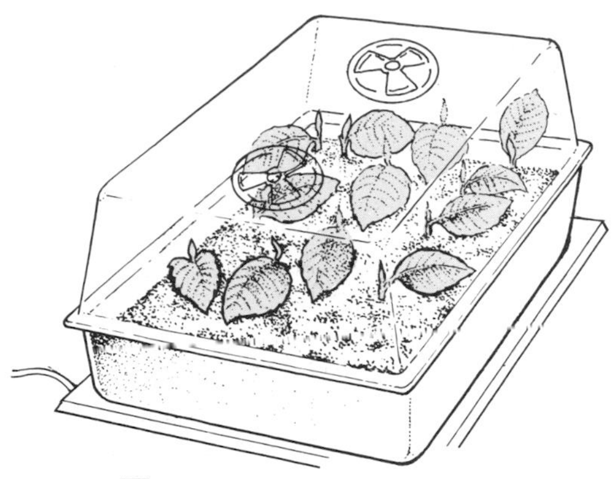

[4] Covering with propagator top.

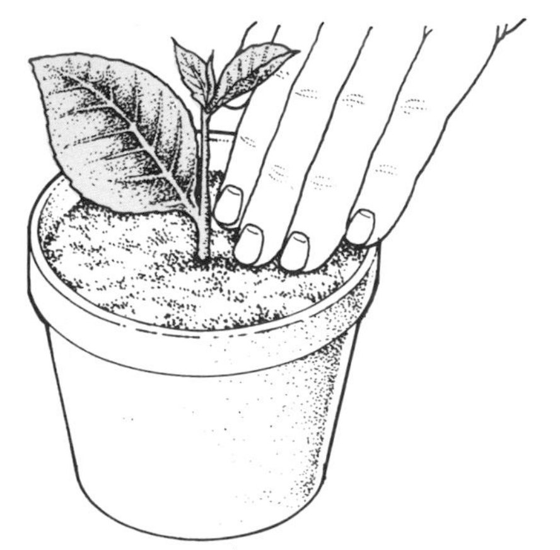

[5] Potting up rooted cutting.

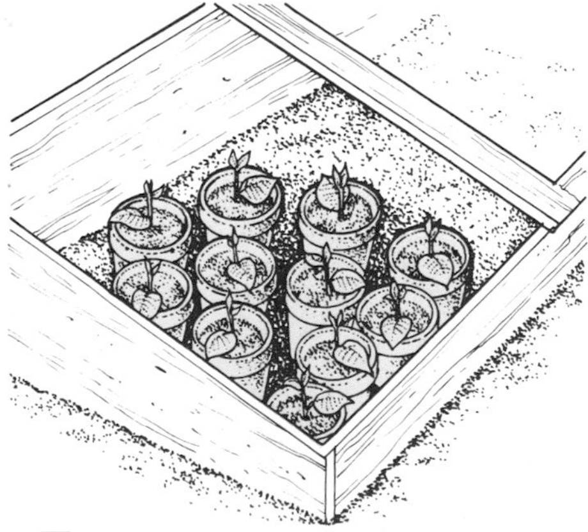

[6] Pots in garden frame to harden off.

[4] Warmth and humidity are particularly important for these rather difficult cuttings. Cover the tray with a plastic top and provide bottom heat in a propagator. Camellias are ideal subjects for a mist unit with soil warming cables if you are lucky enough to have one. Otherwise, mist the cuttings over regularly with water in a hand sprayer.

[5] The cuttings may root within a couple of months, though they may take considerably longer. Don't be in too much of a hurry to move them, but wait until a good fibrous root system has been formed. Water them with a fungicide at the end of the autumn. Whether or not they have rooted, they will need care to see them through the winter. Keep them frost free and quickly remove fallen leaves and other debris.

Once a vigorous root system has developed, pot the camellias individually in an ericaceous (lime-free) compost.

[6] In the spring, move the cuttings to a garden frame and gradually harden them off. With luck, cuttings taken in August should be ready for planting in their permanent positions in the garden in the autumn of the following year.

If your cuttings are very slow to root, they will use up all their stored plant foods and there will be none to replace them in the cuttings compost. If for any reason you leave them in the rooting medium for more than a few weeks, water them with a low-strength solution of liquid fertilizer during the growing season, but not when temperatures are low and daylight hours short.

Vine eyes

Eye cuttings are rather like leaf bud cuttings, but they are taken from fully ripened wood in the dormant season. This method of propagation is most often used for grape vines, but it can be used for many hardwoods; it's particularly useful if propagating material is scarce.

[1] When the leaves have fallen in autumn, cut a sturdy vine stem which has grown the previous year. As always, use healthy, typical growth free of pests. You can generally find some good material for propagation among the prunings; one good branch will provide plenty of cuttings. Check that all the dormant buds are plump and healthy, and the wood is firm. Discard the thinner tip of the branch.

[2] Vine eyes can be taken in two ways. The first is very similar to a leaf bud cutting: make the first cut just above a plump, dormant bud. The second cut is made about 1–1½ in (2.5–4 cm) below.

The second method involves cutting the vine so that the bud is in the centre, with an equal amount of wood each side. Make the cuts with a pair of sharp secateurs, and be careful not to damage the bud.

[3] Use seed and cuttings compost, firmed lightly to just below the rim of the pot. Scatter a thin layer of silver sand on top of the compost to improve the drainage.

Insert the first type of cutting upright, so that the base of the bud is just level with the surface of the compost. The lower cut can first be dusted with a rooting powder formulated for hardwood cuttings.

[1] Cutting length of vine.

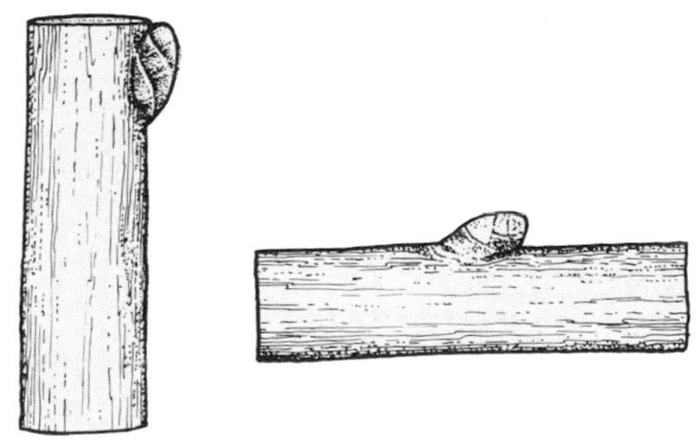

[2] Two types of vine eye.

[3] Inserting upright eye.

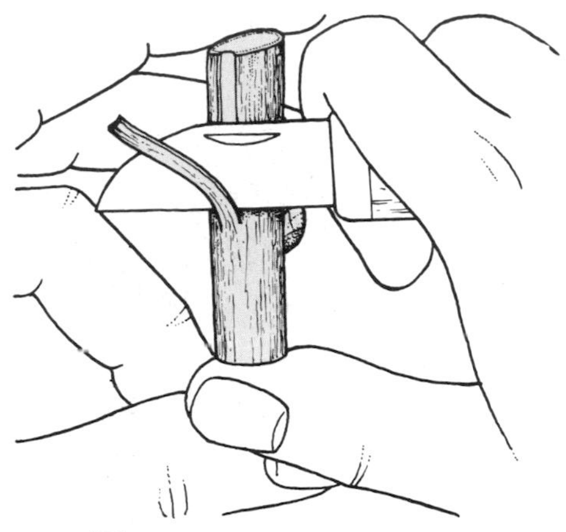

[4] Wounding base of long eye.

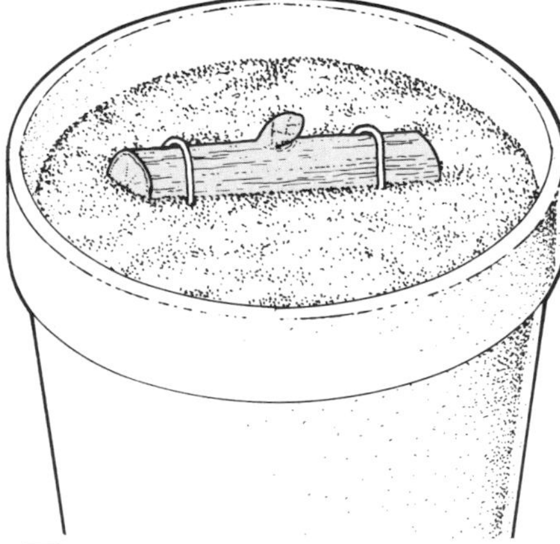

[5] Long eye pegged down in compost.

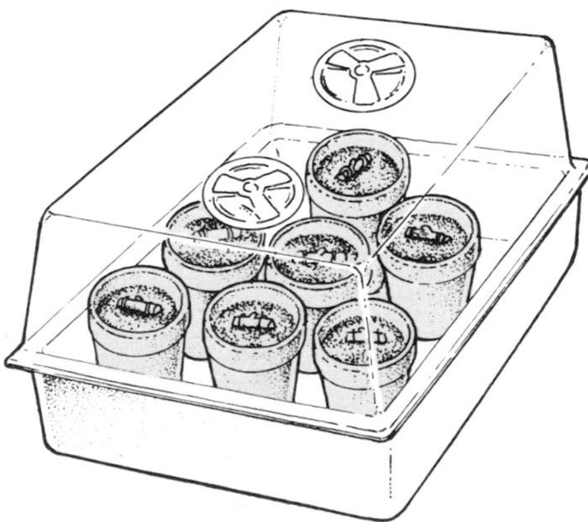

[6] Potted eyes in heated propagator.

[4] The horizontal type cutting tends to root more quickly if a narrow sliver of rind is taken from the side opposite the bud with a sharp knife. This exposes more of the cambium layer from which the roots will spring. Dust this wound, and the ends of the cutting, with rooting powder.

[5] Press the cutting into the compost so that it is half buried, with the dormant bud on top. Hold the cutting in firm contact with the soil by using two pieces of wire bent into hairpin shapes to peg it down into the pot.

Water both types of cutting thoroughly after insertion.

[6] Vine eyes root most rapidly in gentle heat in a propagator, though too much heat will force the buds into growth before there are any roots to support them. Aim to provide a temperature of 13–15°C (55–60°F). A propagator is not essential, but the cuttings must be kept frost free if they are to be successful, unlike other hardwood cuttings, which are happy to be left out in all weathers in the open ground.

In spring, the shoots will start to extend from the buds, and they can be provided with a thin cane for support. Once the cuttings are well rooted they should be moved into 4 in (10 cm) pots of potting compost and hardened off in a cold frame. The following autumn, when the pots are filled with roots, the vines may be planted out in their permanent positions, either in a sheltered spot in the open garden or in a greenhouse, and provided with supports to climb.

Root cuttings

In the same way that stems or leaves can be encouraged to develop roots, some roots can be persuaded to send up shoots, and form new plants that way. A number of garden shrubs and trees can be increased quite easily like this: the amount of shrubs that send up multitudes of suckers show how easy it is for some roots to initiate stems.

[1] Before propagating any plant from root cuttings, make sure that it is not grafted on to a rootstock. Lilacs (*Syringa*) for example, are easily raised by this method, but instead of increasing your stock of a favourite double-flowered white variety, root cuttings may well give you the single, purple-flowered rootstock, *Syringa vulgaris*.

Root cuttings are taken in the dormant season. Choose a day when the soil is workable – not frozen or waterlogged – and avoid days when there are cold, drying winds. Herbaceous plants can be lifted carefully: larger plants, or those that resent disturbance, can have some roots exposed by scraping away the soil.

[2] When the plants have been lifted, wash most of the soil off the roots by gently dunking them in a bucket of water. They don't have to be spotless, just clean enough for you to see what you are doing.

[3] Two or three roots per plant will usually give you sufficient cuttings, and the removal of a small number will have no damaging effect on the mother plant. Make the cuts cleanly with a sharp knife, picking firm, sturdy, undamaged roots. Put them straight into a plastic bag to prevent them from drying out.

As soon as you have taken off as many roots as you need, replant the parent: it wants to be out of the soil for as short a

[1] Lifting plants in dormant season.

[2] Washing roots.

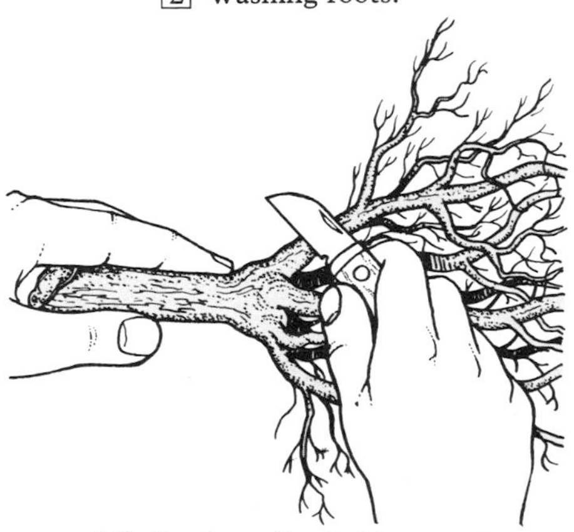

[3] Cutting off suitable roots.

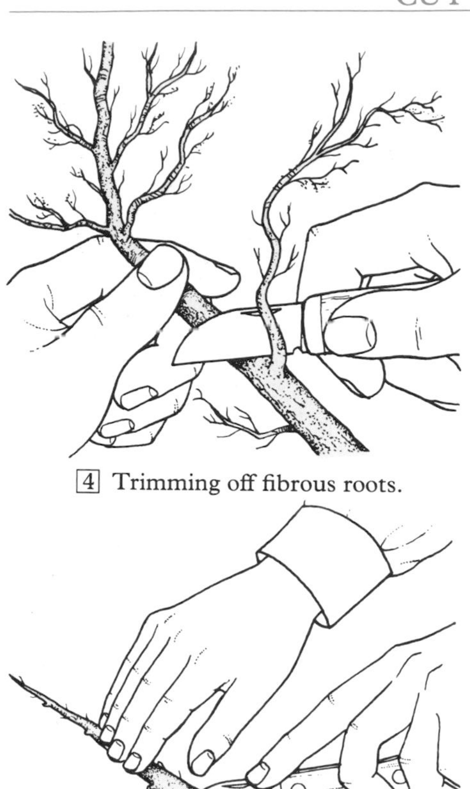

[4] Trimming off fibrous roots.

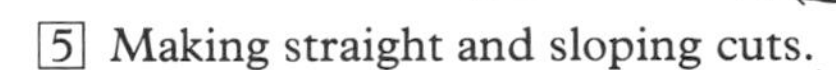

[5] Making straight and sloping cuts.

[6] Watering pot of compost.

time as possible. Fill in round the roots with fine soil and firm the plant in with your knuckles. It is unlikely to need watering unless the soil is exceptionally dry.

[4] Plants which have fairly thick roots are prepared by first trimming off any whiskery side roots with a sharp knife. Try not to make too many cuts and wounds on the root surface as this will increase the likelihood of disease.

[5] The trimmed roots are cut into sections about 2 in (5 cm) long. Before cutting them up, lay the roots out on the bench so that you can tell which is top (that is, the part of the root nearest the parent plant's crown) and which is bottom. Make the top cut straight and the bottom cut at an angle, so that it's easy to tell which way up they should go. If you insert root cuttings upside down by mistake, it won't stop the shoots growing, but they will begin to grow downwards and have to turn themselves round. It slows up the whole process and makes the shoots rather fragile where they bend.

[6] Fill a pot with cuttings compost mixed with a small amount of sharp sand, and firm it down lightly. Water the compost with a fine rose and leave it to drain for a short while before inserting the cuttings.

[7] Dust the cuttings lightly with fungicidal powder. You can either use a brand of fungicide in a puffer pack, or place the cuttings in a plastic bag with a small amount of powder and shake them up together until the roots are lightly coated. *Don't* use a hormone rooting powder on root cuttings. These

powders contain root-promoting hormones; you want shoots. A rooting powder will actually suppress shoot production rather than help it.

Dib a hole in the pot of compost and insert the prepared cuttings until their tops (the straight cut) are just about level with the surface. If you really get confused about which way up the cuttings should go, it's best to lay them on their sides so at least the shoots won't have to do a complete U-turn.

[8] Not all plants have thick, fleshy roots, and thinner rooted subjects are treated slightly differently. The roots are still cut up into 2 in (5 cm) lengths, but without bothering to make sloping cuts to distinguish between top and bottom. Fill a seed tray with cuttings compost and lay the pieces of root on the surface. Cover them with a further, fairly thin layer of compost, about ½ in (1 cm).

[9] Root cuttings don't need heat to start into growth, and can be placed in a sheltered spot outdoors or in a cold frame. The compost should be kept moist, but must be protected from drenching rain. In spring, shoots should appear through the compost, but don't lift the cuttings straight away – shoots often appear before root growth has started. Sometimes the shoots appear first, then new roots spring from the base of the shoot, and not from the original piece of root at all. Once you are sure there is a good fibrous root system, pot the plants up individually and plant them out in the garden the following autumn.

[7] Dibbing hole and inserting cuttings.

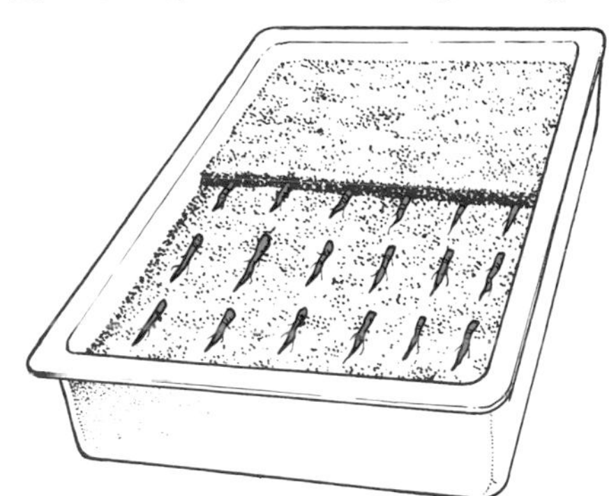

[8] Laying thin cuttings in tray of compost.

[9] Young shoots.

What might go wrong

Leafy cuttings wilt after insertion and fail to recover

•Is the compost sufficiently moist? Water the cuttings thoroughly overhead after insertion. •A humid atmosphere is necessary to prevent leaves and stems from losing all their moisture. Keep propagator lids on, or mist cuttings at regular intervals with plain water. •Too much foliage left on a cutting will lose water faster than it can be replaced. Cuttings with very large leaves can have the leaves rolled or even cut to reduce surface area. •Direct sunshine causes excessively high temperatures. Keep leafy cuttings well shaded until they have recovered from their initial flagging, then put them in a light place, but out of direct sun.

Bases of the cuttings rot below soil.

•Cold, wet soil encourages rotting. Do not overwater the compost. •Ragged cuts, made with a blunt or dirty blade, invite invasion by disease organisms. Use a sharp, clean knife, and prepare cuttings in clean conditions. •Compost for cuttings indoors must be sterile. Once-used compost or garden soil is teeming with potentially harmful micro-organisms (hardwood cuttings are tough enough to stand up to them). •Leaves, leaf stems or stipules buried beneath the soil will provide a site for infection.

Cuttings fail to root and slowly die back from the tip

•Is the type of cutting correct for the type of plant you are trying to propagate? •Does the cutting need bottom heat? •Too high a temperature may force shoots into growth prematurely, and they will die before roots have developed.

Plants which can be increased from cuttings

Softwood cuttings A great number of plants, particularly chrysanthemums, dahlias, fuchsias, pelargoniums, herbaceous border plants, foliage house plants (such as coleus and iresine), and the soft summer growth of a number of shrubs.

Semi-ripe cuttings Most suitable for flowering and foliage shrubs, both deciduous and evergreen: *Abelia*, *Berberis*, *Choisya*, *Cotoneaster*, *Daphne*, *Erica*, *Escallonia*, *Garrya*, *Hebe*, *Jasminum*, *Kerria*, *Lavandula*, *Mahonia*, *Parthenocissus*, *Philadelphus*, *Rosmarinus*, *Sambucus* and *Viburnum* are a few examples.

Hardwood cuttings Trees and shrubs such as *Buddleia*, *Cydonia*, *Deutzia*, *Forsythia*, *Ligustrum*, *Lonicera*, shrubby paeonies, *Populus*, *Ribes*, *Salix*, *Symphoricarpos*, *Weigela*, *Wisteria*: and fruit bushes such as black, red and white currants and gooseberries.

Leaf cuttings Whole leaves: *Saintpaulia*, *Begonia boweri*, *Peperomia*; Strips: *Streptocarpus*, Gloxinia, *Sansevieria*; Pegged down leaf with slit veins: *Begonia rex*.

Leaf bud cuttings *Camellia*, *Mahonia*, *Ficus elastica*.

Vine eyes *Vitis* (grape vines); also any plant which can be propagated by hardwood cuttings.

Root cuttings Herbaceous plants such as *Acanthus*, *Anchusa*, *Anemone* × *hybrida*, *Anemone pulsatilla*, *Brunnera*, *Papaver*, *Phlox*, *Primula denticulata*, *Romneya*, *Verbascum*; also any shrub or tree that produces suckers should grow from root cuttings: *Ailanthus*, *Daphne genkwa*, *Daphne mezereum*, *Prunus* (ornamental plum), *Rhus typhina*, *Robinia pseudoacacia* and *Syringa*, plus climbers *Campsis radicans* and *Eccremocarpus scaber* and one vegetable – seakale (*Crambe maritima*).

3
LAYERING

Equipment

Many plants propagate themselves by layering – some do it all too frequently! – so obviously not much special equipment is necessary.

The trusty sharp knife (*a*) is needed to slit the stems of shrubby branches, and a trowel (*b*) for digging the hole to receive the layer. Springy branches will have to be pegged down firmly to keep them in place, and wire (*c*) bent to the shape of a hairpin does the job very well. Choose wire of a slightly thinner gauge than that used for coathangers – such wire is easy to bend while still strong enough to do the job properly.

Many layers are pegged down direct into the garden soil surrounding the shrub, while others can be rooted into pots of various shapes and sizes (*d*). This method has the advantage that the new plant can easily be lifted without any root disturbance.

Once rooted, the layer must be separated from the parent plant before it can be lifted and replanted; make a clean cut with a pair of secateurs (*e*).

Air layering (usually performed on indoor plants) requires the use of a plastic sleeve to hold the rooting medium in place. Black polythene is sometimes recommended, but clear polythene lets you see when the layer has made plenty of roots, and has no detrimental effect on rooting. Self-clinging food wrap (*f*) works very well.

If you are to get a good, straight stemmed plant from a layer, the tip should be tied upright to a cane (*g*) as soon as the layer has been made.

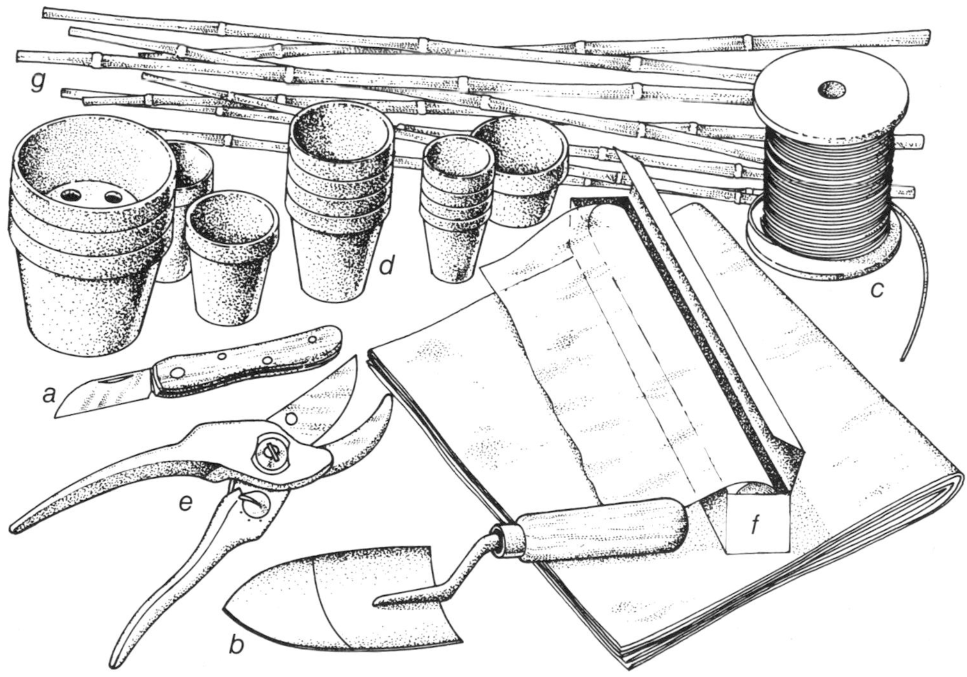

Layering equipment: for details see text above.

[1] Selecting suitable shoot.

[2] Bending shoot to soil.

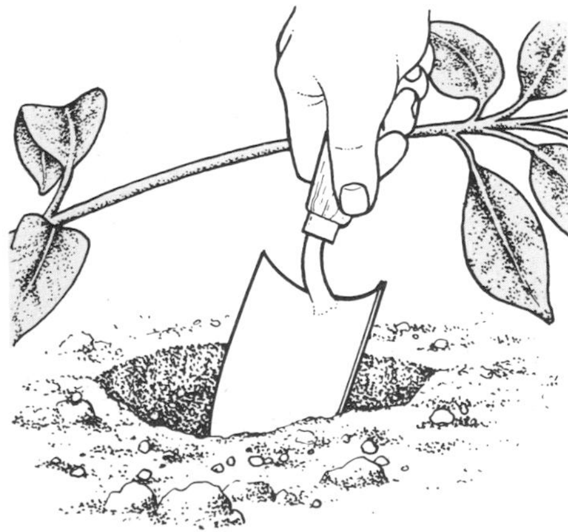

[3] Making hole with trowel.

Blueprint for layering

Rhododendrons, difficult to propagate by other methods, lend themselves to layering, requiring only a small amount of patience – the layers may need to be left *in situ* for 18 months before they are ready to be planted out.

[1] In early spring, choose a healthy, vigorous shoot that grew the previous year and can be bent to the ground easily. Sometimes a plant will have no suitable bendy branches. In this case, cut one or two of the lower branches back hard: the following season they should produce vigorous extension growths that will supply you with the shoots you need for layering next year. (If you can't wait that long, you could air layer a branch instead.)

Dig over the soil around the plant thoroughly, but be careful not to damage the plant's roots.

[2] Grasp the chosen shoot about 9–12 in (23–30 cm) behind its tip and bend it down to the soil. Mark the position of the stem on the ground at the point where you are holding it, using a small stick, large stone or something similar as a marker. This is the point where you will bury the shoot. Release the stem while you prepare the soil.

[3] Using a trowel, dig a hole about 4 in (10 cm) deep at the marked point, making the side furthest from the plant as straight as possible. Slope the base of the hole upwards, back towards the plant, until it makes a trench 8–10 in (20–25 cm) long to receive the stem. If the soil is poor or badly drained, work in some peat and sharp sand. Lining the hole with peat will help speed up rooting.

[4] In layering, a branch is induced to produce adventitious roots (they spring directly from the stem) even though it is still attached to the parent plant. This is achieved by causing a 'traffic jam' of plant foods and hormones at a certain point in the stem, and keeping the stem dark.

The leaves and leaf stems are trimmed off the branch to be layered for a length of about 2 ft (60 cm), starting 6 in (15 cm) behind the tip. When the branch is pulled down into the hole, it will be bent at a right angle, which in itself constricts the flow of hormones and nutrients. To make doubly sure, the stem is usually wounded at this point, too. The simplest way is to make an angled cut with a sharp knife, cutting less than half way into the stem to avoid it snapping entirely. A small stone or splinter of wood can be pushed into the cut to prevent it from healing, if you like. Roots will now spring from the stem immediately above the wound and the sharp bend.

[5] Pull the prepared shoot down into the hole so that the leafy tip is above the ground. Using a piece of bent wire, peg the stem firmly into the soil. Pliable growths tend to spring up out of the soil unless they are pinned down, even when they are buried.

[6] If the excavated soil is heavy, mix it with some peat and sand before returning it to the trench. Darkness is necessary for roots to be produced, so the whole stem must be thoroughly covered. Tread the soil down lightly to firm it, then water it well.

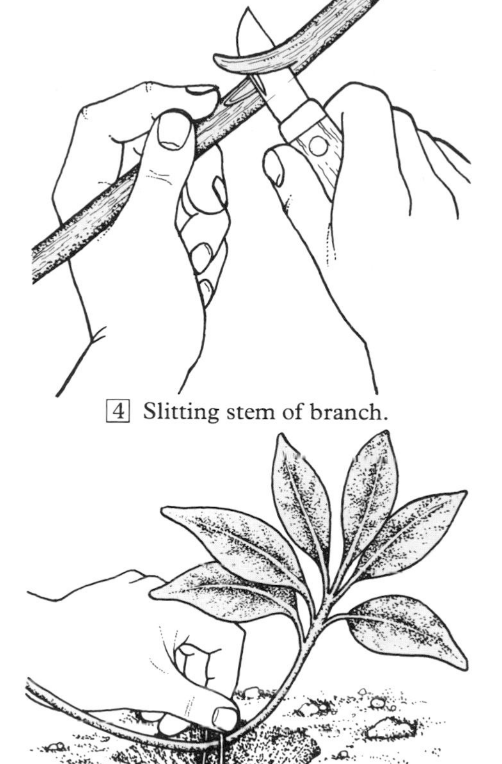

[4] Slitting stem of branch.

[5] Pegging shoot into soil.

[6] Mounding soil over layered shoot.

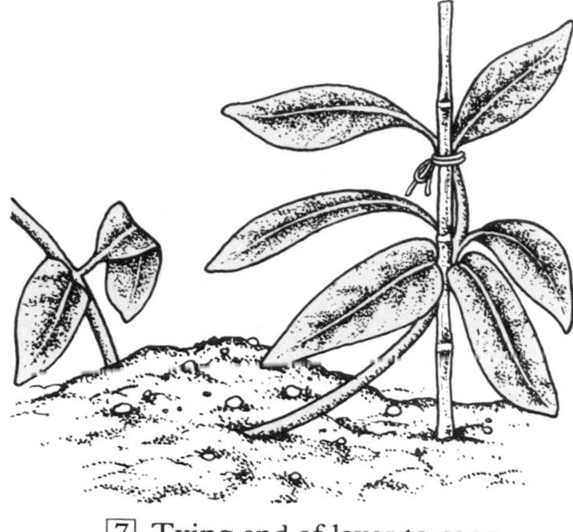

[7] Tying end of layer to cane.

[8] Severing layer from parent.

[9] Digging up rooted layer.

[7] The tip of the layer which is showing above the soil should be kept straight if a shapely plant is to be produced, but will have a tendency to lean away from the parent plant. Insert a cane by the tip and tie the shoot to the cane as it grows.

[8] Roots should be produced during the following growing season. The soil around the layered stem must never be allowed to dry out, and you will need to water it frequently during the summer.

Because the branch is still attached to the parent it is still receiving a supply of food and water, even though this supply has been slowed down. Severed stems (cuttings) have to rely entirely on their own stored food supply, and if this runs out before the roots have formed, the cutting will die. Layering is therefore a much more successful way of increasing difficult subjects.

By the autumn following spring layering, the new plant should be sufficiently well developed to be separated from its parent.

[9] Don't dig the layer up straight away, but leave it to get established on its own for at least six weeks. After this, a well rooted layer can be lifted and moved to its permanent site.

If the layer has not developed a good, fibrous root system it should be left undisturbed for a further year. (The state of the roots can be assessed to some extent by the amount of growth the top has made.) In either case, pinch out the top 2 in (5 cm) or so of the new plant early the following spring; this will encourage both branching and good root growth.

Serpentine layering
This form of layering produces a number of plants from one shoot, which must be long and pliable.

[1] Take a vigorous, healthy young clematis shoot down from its support in spring. Parts of the stem will be buried, and roots will be produced from the buried node (where the leaves join the stem). For each buried node, a length of stem with at least one plump bud or strong shoot must be left above ground, as new shoots will not spring from the buried portion.

Choose suitably spaced nodes and remove the leaves from around them, being careful to leave two or three good shoots between each prepared portion.

[2] With a sharp knife or razor blade, make a small, slanting cut immediately behind each prepared node to wound the stem and encourage rooting. Prepare the soil in front of the plant by digging it over and adding peat. Lay the prepared shoot on the ground (it can be curved round to take up less space) and with a trowel, dig a shallow hole by each prepared node. Lay the stems in the holes and bury them firmly.

[3] When the layer is completed, it will have a series of loops of stem above ground – so it looks like a snake or caterpillar making its way through the soil! Good root systems should have formed by autumn, when the whole layered stem can be lifted and cut into portions, each one having roots and a length of stem with buds and shoots.

[1] Leaf removal from suitable shoot.

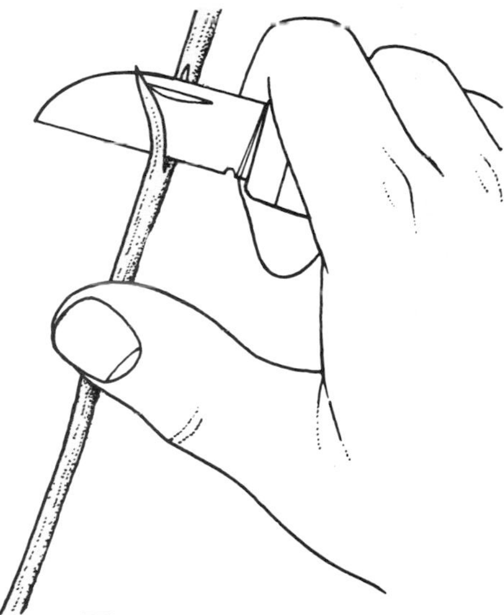

[2] Wounding stem at node.

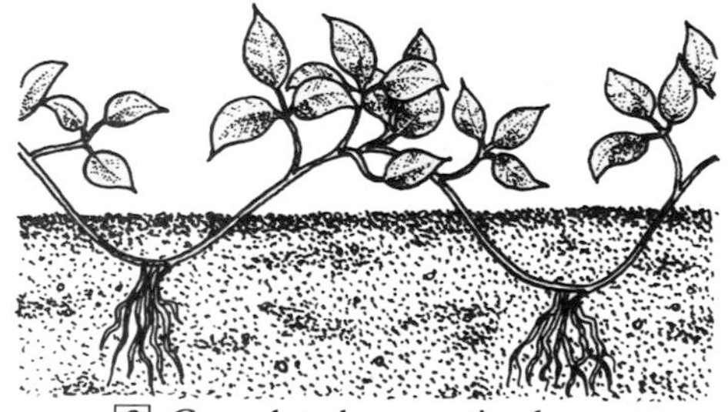

[3] Completed serpentine layer.

Tip layering

[1] Members of the blackberry family produce a new plant when the complete tip of a young stem is buried in the soil. Wild brambles do this with ease, which is why they spread so quickly, forming tough loops of arching stem, firmly anchored at both ends. Efficient snares for unwary walkers!

Take a supple stem of the current season's growth in July, with a strong, pest- and disease-free growing point.

[2] Bend the stem to the ground, and where the tip touches the soil, dig a hole about 4 in (10 cm) deep with a trowel. Heavy soils should be improved with a little peat and sharp sand.

[3] Pull the tip into the hole and bury it, firming the soil round it well. In light soils the flexible stems may spring out of the ground, but they can be anchored with a piece of bent wire. Water the layered tip well, and keep the soil moist during the following weeks.

[4] During August a new shoot will appear from the buried tip and will grow away strongly. The soft new growth will be attractive to greenfly, and as these spread virus diseases, it's worth spraying the new plant with a contact insecticide to keep it greenfly and virus free.

In early autumn, cut the layered cane close to the parent plant, and again about 10 in (25 cm) from the rooted layer – the portion next to the layer is useful for handling the new and rather fragile plant, which can be moved to its permanent position later in the autumn.

[1] Selecting suitable tip.

[2] Making hole with trowel.

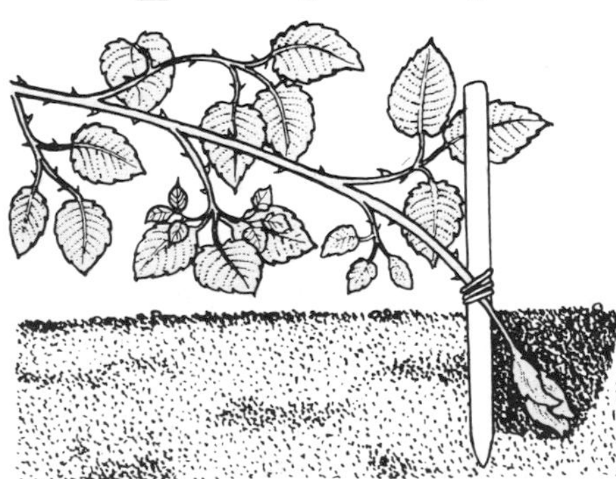

[3] Buried tip.

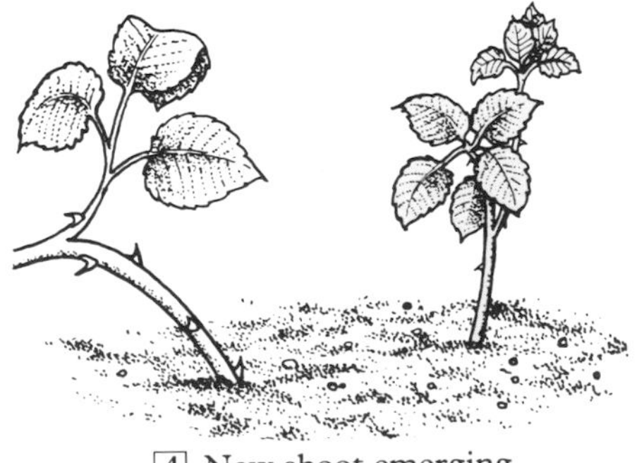

[4] New shoot emerging.

Runners

Strawberry plants begin to deteriorate after a few years, and a bed should be renewed every third year for the best crops. It's easy to produce replacement plants: if one-third of the bed is replanted each year you'll never be without plenty of strawberries.

[1] Strawberries produce new plants on runners in such abundance they become a nuisance if you don't want to propagate them. In August, select four or five well-spaced runners on each of the healthiest, strongest plants, and cut the rest off close to the parent.

[2] Strawberry runners can be rooted direct into the soil surrounding the plant; they certainly establish themselves there with ease if left to their own devices. However, it's worth sinking a pot (preferably clay, though a plastic one will do) into the soil and pegging the runner into that; it makes lifting the new plant easier and avoids damaging the fibrous root system. Fill the pots with good garden soil or a loam-based potting compost.

[3] Each runner carries two or three plantlets, getting smaller towards the tip. The one nearest the parent plant is the strongest, and the rest of the runner beyond this should be cut off with a sharp knife. If you look closely at the plantlet you will probably find that it already has the beginnings of a vigorous root system.

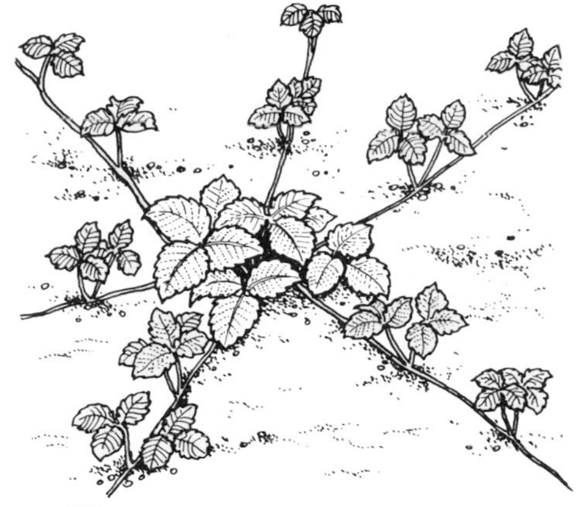

[1] Strawberry plant with runners.

[2] Sinking pot into soil.

[3] Cutting end of runner beyond first plantlet.

[4] Pegging plantlets into pots.

[5] Severing runner stem from parent plant.

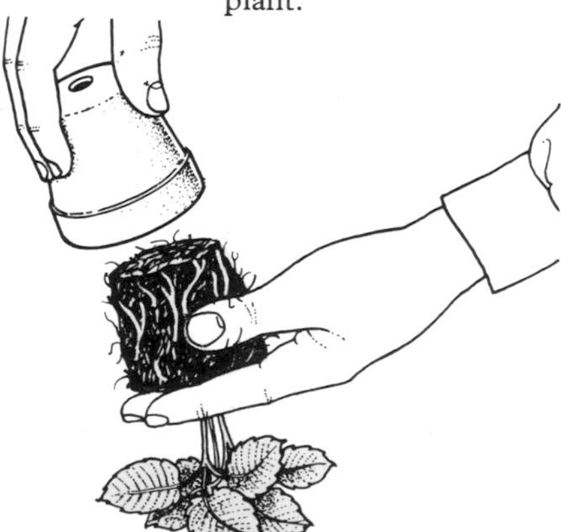

[6] Removing plantlet from pot.

[4] Peg the plantlet down into the pot so that it is making good contact with the soil; use a wire staple or large pebble on the runner to hold it in place. Water the pot thoroughly, and make sure the runners are kept moist during the next few weeks. Clay pots are porous, and will be able to absorb water from the surrounding soil as long as there is sufficient rain: the soil in plastic pots is more likely to dry out, and will need checking.

[5] It won't be long before the plantlet on the runner has pushed out new roots and established itself firmly in the pot. It should form a good leafy crown, and may even be enthusiastic enough to throw out some runners itself – these should be cut off. Like the blackberry family, strawberries are prone to virus diseases which are spread by aphids. It's worth spraying the whole bed with a contact insecticide as soon as greenfly are noticed, before they have a chance to build up.

Once the plant is well established, cut the runner stem close to its crown to sever it from the parent plant.

[6] Late September is the best time for planting. Prepare thoroughly the bed where the new plants are to go, removing all perennial weed roots and incorporating a long-lasting fertilizer such as bonemeal.

Lift the complete pot containing its rooted runner, invert it and tap the rim on a firm surface to loosen the plant. It should turn easily out of the pot with a good rootball. Dig a hole with a trowel, and firm the plant in well with your knuckles.

Air layering

[1] Air layering is carried out on plants that cannot be bent down to the soil; instead, you have to bring the soil up to them. Although outdoor shrubs like rhododendrons can be air layered, the most frequent candidate is indoors: the rubber plant.

After several years, rubber plants usually become lanky and ungraceful – a tall bare stem with leaves only at the top. If you induce roots to grow just below the lowest leaves, you can cut off the top and have an attractive plant again.

Make a wound in the stem about 4–6 in (10–15 cm) below the lowest leaf, making an oblique cut with a sharp knife. Cut less than half way through the stem. Dust a small amount of rooting powder into the cut with a soft brush, and wedge the tongue open with a matchstick or a small wad of moss.

[2] Moist sphagnum moss is the ideal rooting medium, and should be packed right round the stem in the region of the cut. If pressed lightly together it moulds itself well. If you can't get hold of any moss, coarse peat or peat-based cuttings compost can be used instead, but in this case you will have to fix the polythene sleeve in position first.

[3] Wrap a piece of clear polythene round the sphagnum moss. The self-clinging plastic film sold as kitchen food wrap is easy to handle and does the job well. Tape round the top and bottom of the plastic to attach it to the stem and hold the whole thing in place.

If you are using peat instead of moss, cut the polythene to size and tape it in place round the bottom and along the overlap so that it forms a funnel shape. Fill it up with moist peat and tape the top as before.

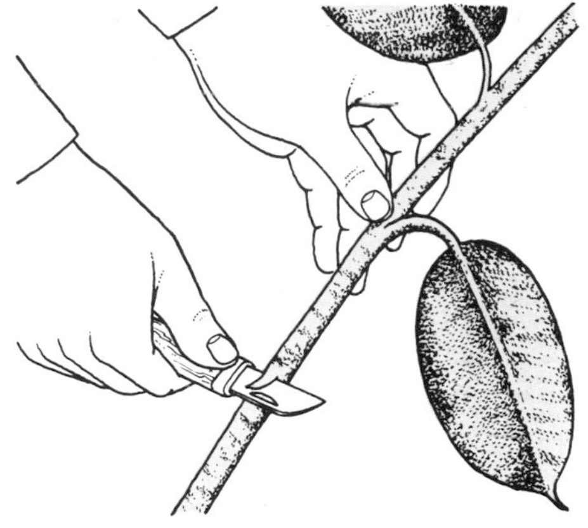

[1] Wounding stem.

[2] Packing with moss.

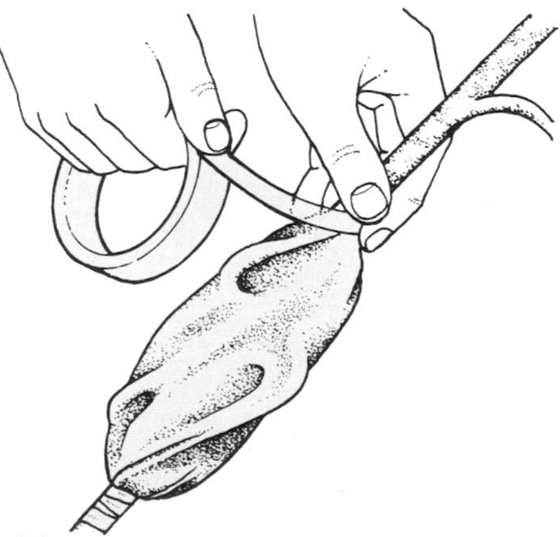

[3] Taping cling film at top and bottom.

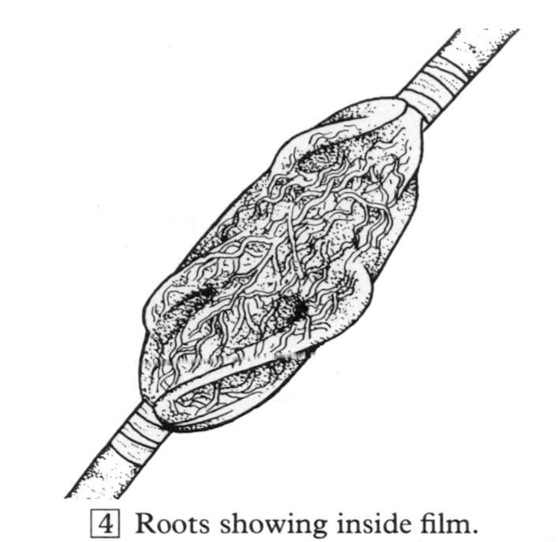

[4] Roots showing inside film.

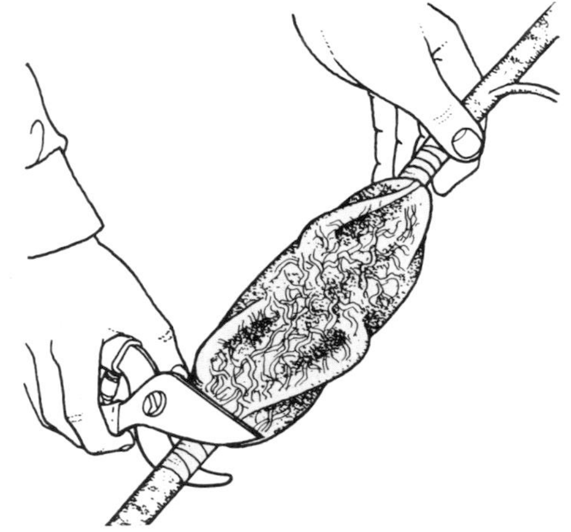

[5] Severing rooted layer.

[6] Planting new layer.

[4] Keep the plant well watered and in an even temperature. Air layering is a slow process and it will be a matter of months before you see anything happening, but the certain sign of success is when a network of fine white roots becomes visible amongst the sphagnum moss.

[5] Once plenty of roots are in evidence, the top of the plant can be cut off with secateurs just below the plastic film. Slit the film open with a knife or razor blade, and carefully peel away most of the moss without damaging the roots. If peat or compost has been used, just shake off the excess.

The original rubber plant needn't be discarded. Cut the stump down to about 9 in (23 cm), just above a bud, and keep the plant warm and moist. It should start into growth and produce several shoots, making a bushy specimen which is, if anything, more attractive than the original.

[6] Pot the rooted layer into a 5 in (13 cm) pot, using a peat-based potting compost. Grow it on until you can start the whole process again!

What might go wrong

The layer doesn't root

•Make sure the layer is in good, firm contact with the soil, held down with a peg or stone. Pliable branches can spring out of the soil some time after layering if not secured. •Wound the branch in the correct place; although sometimes layers will root without wounding, where recommended it improves the chances of success. •Wounding should be done cleanly. Hacking at the stem may encourage rots. •If you are too enthusiastic about wounding, you risk the branch snapping off altogether. Never make the cut more than half way through. •Check you have allowed sufficient time for roots to form. Sometimes layering is a slow process – don't be too impatient!

The rooted layer dies when separated from the parent and transplanted

•Rooted layers are often quite large plants, and the larger the plant, the more difficult it is to establish successfully. Treat it with care; the new roots are fragile and easily damaged. Dig the plant up carefully, retaining plenty of soil around the root ball. •Transplants with a lot of foliage lose water quickly and wilt easily. Protect the newly planted layer from drying winds and shade it from the sun. Water after transplanting.

The new plant is unhealthy

•Any disease the parent has will be passed on to the layer. Strawberries, raspberries and blackberries in particular suffer from a range of virus diseases. Ensure the parent stock is healthy before propagating from it.

Plants suitable for layering

Blueprint method (simple layering) Particularly useful for magnolias and rhododendrons, which are difficult to increase by other means. Other shrubs which can be layered include *Forsythia suspensa*, various ornamental *Prunus* species, *Cornus* and *Chaenomeles*, though almost any shrub will respond to this method as long as the branches can easily be lowered to the soil. Plants with an arching or weeping habit are especially suitable.

Border carnations and pinks can also be layered, as can the foliage houseplant dieffenbachia, which often loses its lower leaves and develops bare, leggy stems.

Serpentine layering Plants with pliable, climbing stems, such as *Aristolochia*, *Clematis* (species and large-flowered hybrids), *Jasminum*, *Lapageria rosea*, *Lonicera*, *Parthenocissus*, *Vitus* (grape and ornamental vines) and *Wisteria sinensis*.

Tip layering Members of the blackberry family, including loganberries and boysenberries: also ornamental species such as *Rubus ulmifolius*.

Runners Strawberries (large-fruited types and some alpine varieties – not all alpine strawberries will develop runners), *Ajuga reptans*, *Chlorophytum*, *Geum*, *Saxifraga sarmentosa*, *Viola odorata* and its varieties.

Air layering Any woody shrub which would normally be simply layered but which may have no suitable branches for bending to the ground, particularly rhododendrons and magnolias. Most commonly used on houseplants which have lost their lower leaves and developed ugly, bare stems, including *Ficus elastica*, *Dracaena* and *Citrus* species.

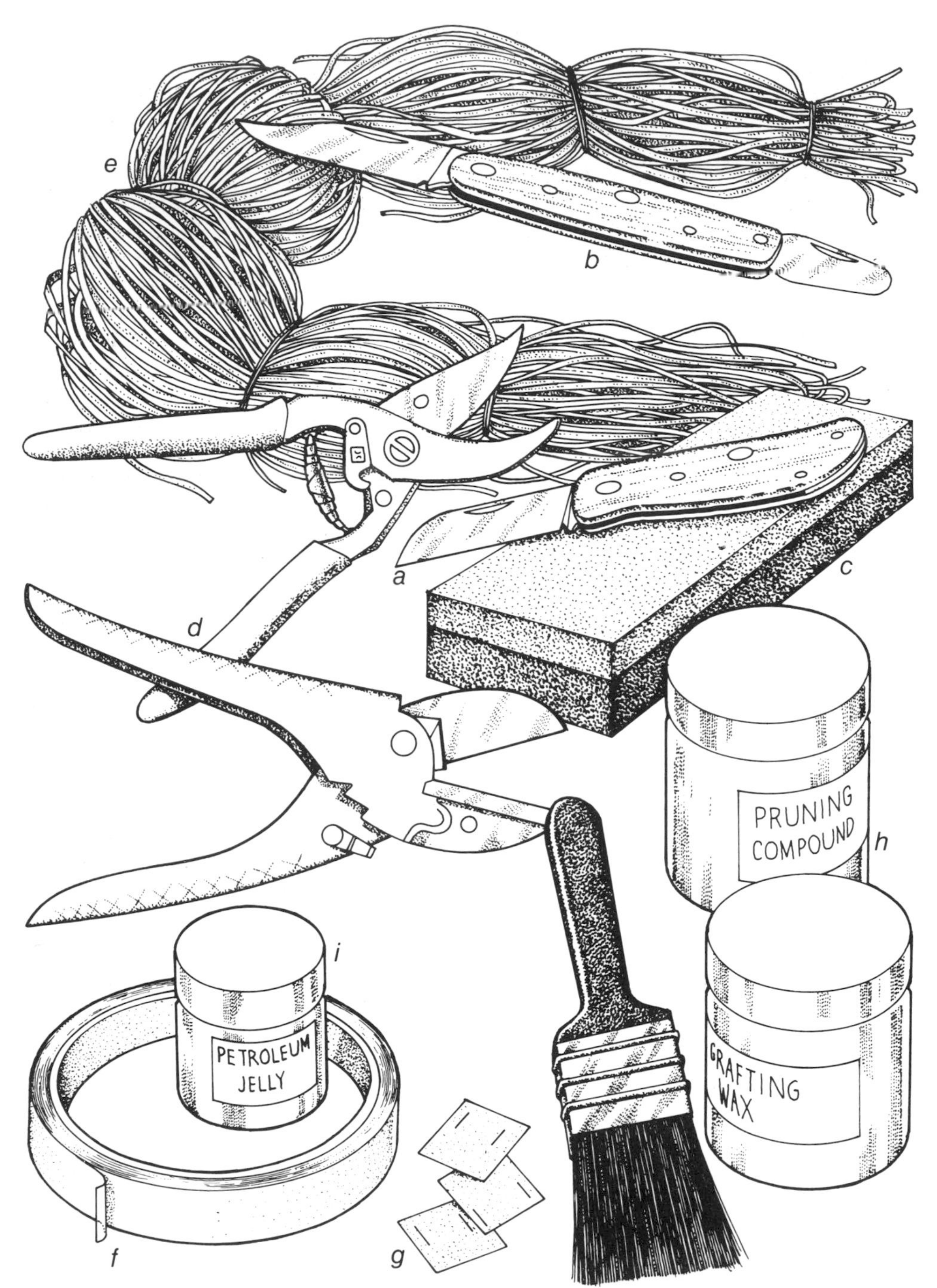

Grafting equipment: for details see text opposite.

4
GRAFTING

Equipment

Budding and grafting probably demands more skill than any other method of propagation. It is not difficult, but it does need care, practice and a good set of tools.

I have talked a lot about using a sharp knife already, but nowhere is it more vital than in grafting. Choose your knife with care. It must be comfortable to hold, preferably with a smooth handle, sturdy and reasonably heavy. A grafting knife (*a*) must have a straight blade. Avoid stainless steel blades – they don't hold their edge as well as mild steel. Also avoid knives with curved blades; although often sold as grafting knives they are actually for pruning and are useless for grafting.

A budding knife (*b*) is not vital if you have a grafting knife, but it does make budding a bit easier. A slightly smaller version of a grafting knife, the main difference is in the end of the handle (or sometimes the blade) which is flattened to a broad, spatulate shape and made of bone or plastic. It's used to prise the flaps of bark away from the stem when budding. You can use the blade of the grafting knife instead, or you can use the budding knife for grafting cuts, so only one of these knives is essential.

To keep your knives super sharp, you'll need a sharpening stone (*c*). They are available in a range of sizes, but you'll get the best results from a large stone that can be fixed to the bench in a block of wood.

Secateurs (*d*) will be needed for cutting your supplies of grafting material, and a strong, sharp pair will make sure you don't damage the trees and shrubs. Parrot-billed or anvil secateurs are equally suitable.

The two portions of plants which are to be grafted have to be held together firmly, and while your carpentry will (of course!) ensure a good fit, some extra precautions are necessary. Raffia (*e*) is the traditional material for tying, and is very good as long as it's soaked before use. Polythene tape (*f*) is often used now instead, as it's a little easier to handle than a hank of raffia. The non-sticky sort isn't easy for gardeners to obtain, but transparent sticky tape does work if you bind it tightly (remember that it won't stay sticky once wet).

Budding patches (*g*) are small squares of rubber with a metal staple, and provide a quick and easy way of securing the bud to the rootstock. Stretched right over the bud and fastened at the back, they will have perished by the time the bud is ready to start growing.

Keep air and water out of grafts by using wax or a pruning paint (*h*), daubed over the whole area with a small brush. Hot wax is the traditional method, but just as effective and much easier to use is one of the proprietary bitumen-based sealing compounds. These are not suitable for use on the tender, easily damaged buds used in budding, but pure petroleum jelly (*i*) smeared over the bud protects it and increases the chances of a 'take'.

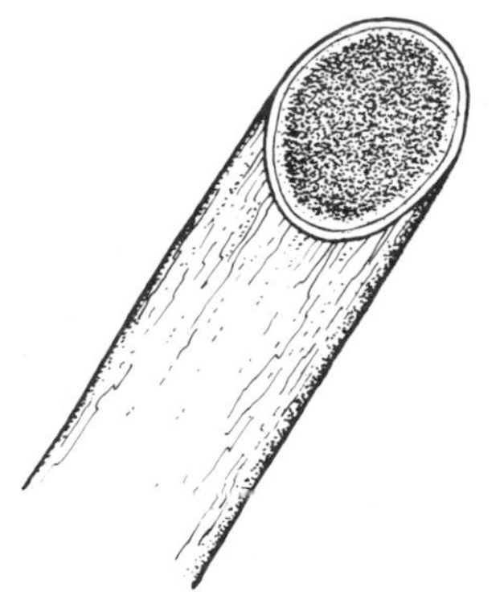

[1] Stem section showing cambium layer.

[2] Selecting suitable stems.

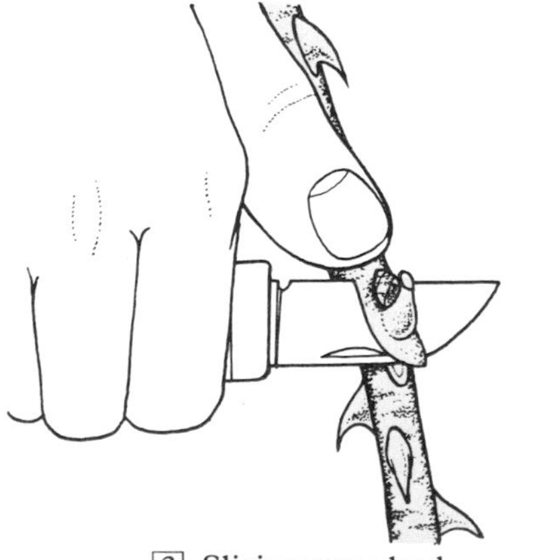

[3] Slicing out a bud.

Blueprint for shield budding

Grafting is joining parts of two separate plants together so that they will continue their growth as one plant. Budding is just one method of grafting.

Plants are usually grafted so that the roots are of one variety, while the top growth – the scion – is different. Often the rootstock has a beneficial effect and is an improvement on the scion's natural roots. Sometimes, as with roses, plants are rather difficult to propagate by other methods, and budding gives the quickest, most reliable results.

[1] Much of any plant's stem consists of dead, woody tissues which are only there for support. The living cells exist in a thin ring between the central pith and the bark; this ring is called the cambium layer. If you cut across a branch you will be able to see this layer as a thin, green line. When grafting, it's essential to get the cambium layers of both plants into contact, or the tissues will not fuse.

[2] Once you have established suitable rootstocks, the first step in budding is to collect material of the variety you want to propagate. This is done in summer, when flowering has just finished, cutting firm, healthy stems that have grown that season. Remove the leaves from the stems.

[3] On warm days, keep the budsticks you have just prepared in a plastic bag while you are working. Standing by the rootstock, cut out a suitable bud with your knife, starting just below it and scooping it out shallowly on a shield-shaped piece of rind.

4 Once you have cut the bud, keep it moist and clean while you deal with the rootstock. You can drop the bud into a dish of water – or put it on to your tongue if you want to be truly traditional!

The rootstocks should be well established plants with a clear stem at the base. You can either raise rootstocks yourself or buy them in from specialist rose nurseries the preceding autumn. Keep the rootstocks well watered, especially the week before you want to bud them. This makes sure the rind will lift easily from the wood.

You have exposed the cambium layer on the bud: now you need to do the same on the rootstock. Make a shallow, T-shaped cut on the stock 1–2 in (3–5 cm) from the ground. This is most easily done by bending the stock back so the rind you are cutting is under tension.

5 Slip the handle of your budding knife into the T-cut you have just made so that it raises the two flaps of rind from the wood. You should be able to see the cambium layer as moist, green tissue just under the rind.

6 Holding the bud by its little 'tail' of rind, push it down into the T-cut on the stock until it fits snugly. Be careful not to damage the bud while you're doing this. The two layers of cambium will now be held firmly in contact by the two flaps of rind, but the bud will need securing in some way to make sure it doesn't get knocked out.

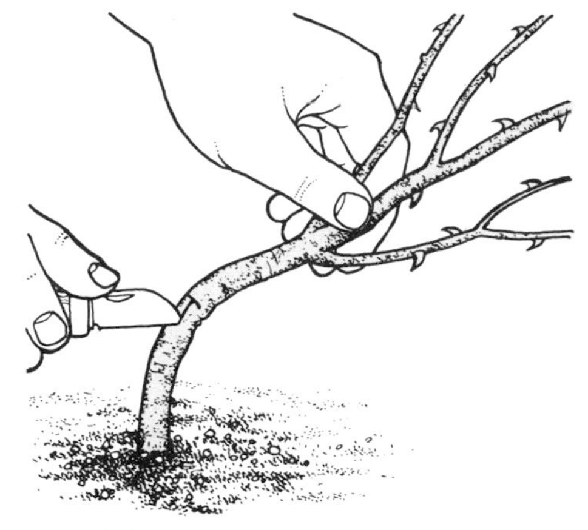

4 Making a T cut on stock.

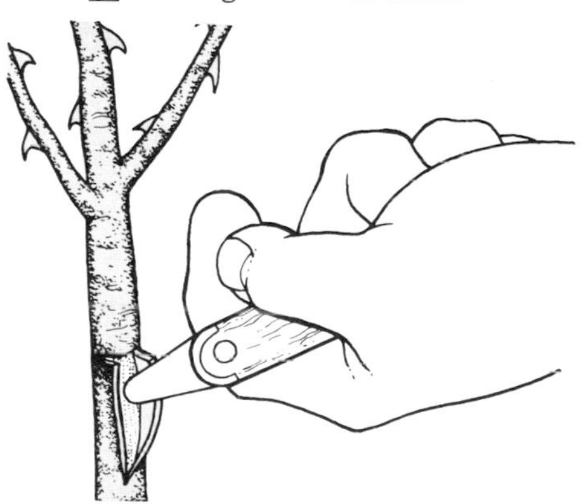

5 Lifting rind with end of budding knife.

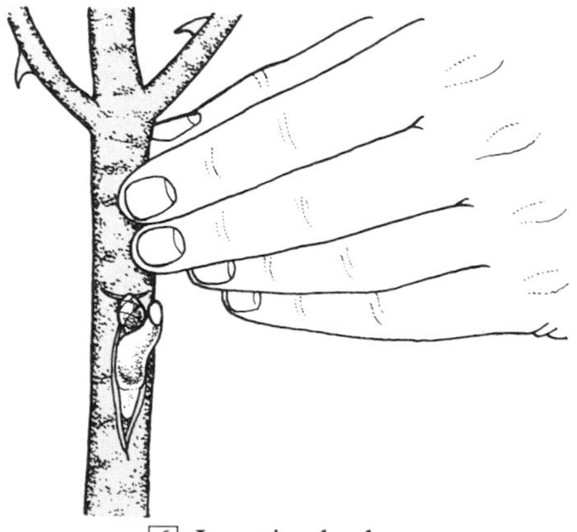

6 Inserting bud.

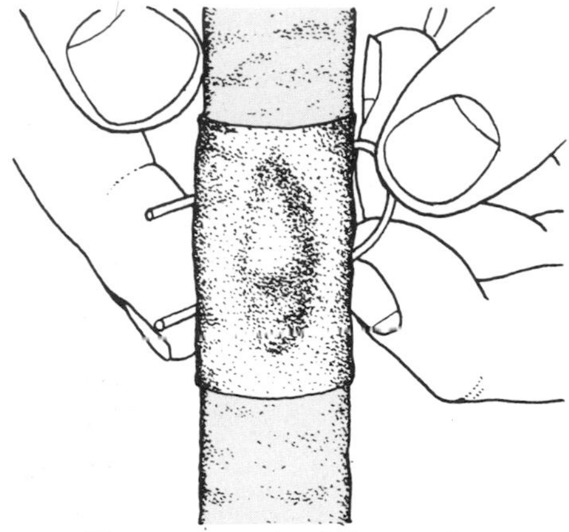

[7] Putting on budding patch.

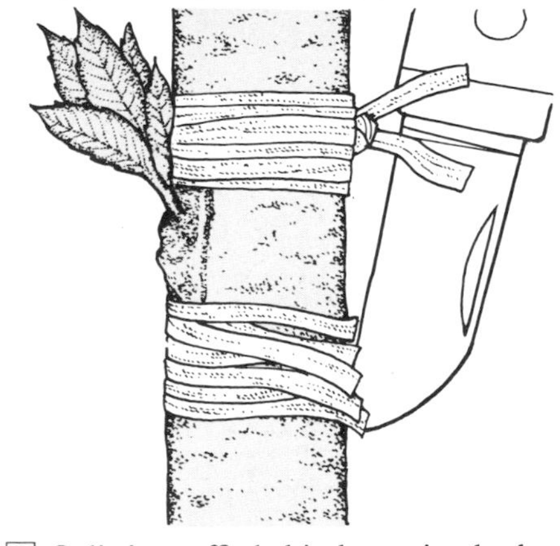

[8] Splitting raffia behind growing bud.

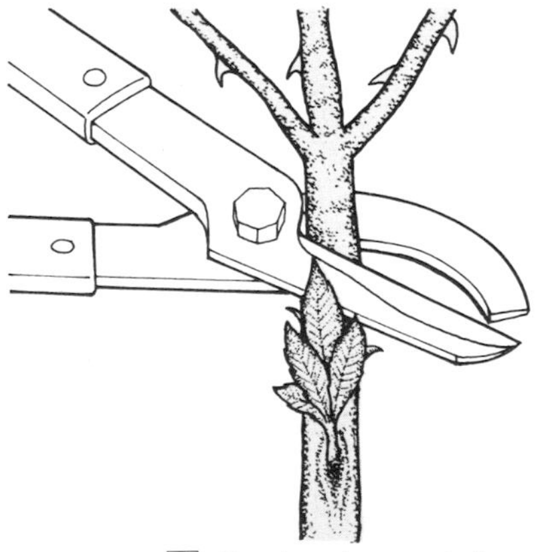

[9] Cutting down stock.

[7] Raffia can be used to hold the bud, but a budding patch is a simpler method. This thin piece of rubber is stretched over the bud until the two ends meet round the back of the rootstock's stem; they are held together there by the metal staple. As rubber is perishable it will have disintegrated within the time it takes for the bud to start growing, so it won't impede its development in any way, but it will keep the bud and rootstock in good close contact while they unite.

[8] Depending on the season, it will take about a month for the bud and stock to fuse. The bud may start to grow away in the autumn following budding, or it may not make a move until the next spring. When it does start to grow, if you have used raffia to tie the bud into place it must be split carefully up the back or it will restrict the bud's growth.

[9] Leave the budded stock intact through the autumn and winter. In very early spring, before growth starts, cut off the top growth of the rootstock just above the bud. The budded variety will then grow out as the only shoot, so it will form all the top growth of the plant.

Keep the new plant weed free and watered through the spring and summer. The following autumn it may be lifted and transplanted to its permanent position in the garden.

Chip budding

Shield budding must be done while the plants are in growth, when the rind lifts easily from the wood, while chip budding can be done while the plants are still dormant. Good results are obtained from chip budding in early spring, just before growth starts: it can be performed during the growing season, too. It is not as easy as shield budding, however, and it's generally only used when shield budding is impossible.

[1] Collect the scion wood (the variety which is to be grafted on to the rootstock) choosing firm, plump, healthy shoots with well developed buds. Remove leaves and leaf stalks if these are present.

[2] Cut the bud out with two actions. First, make a downward sloping cut into the stem, beginning about $\frac{1}{4}$in (5 mm) below the base of the bud. Withdraw the knife and make the second cut beginning about $\frac{1}{2}$in (1 cm) above the bud, and slicing shallowly behind it until the knife meets the first cut. This will remove the bud on an arch-shaped piece of rind with a little wedge at the base.

[3] Now make a cut on the rootstock in just the same way, so that the bud will fit *exactly* on to the stock. The matching of the two cut portions must be perfect, and you might need one or two attempts to achieve this!

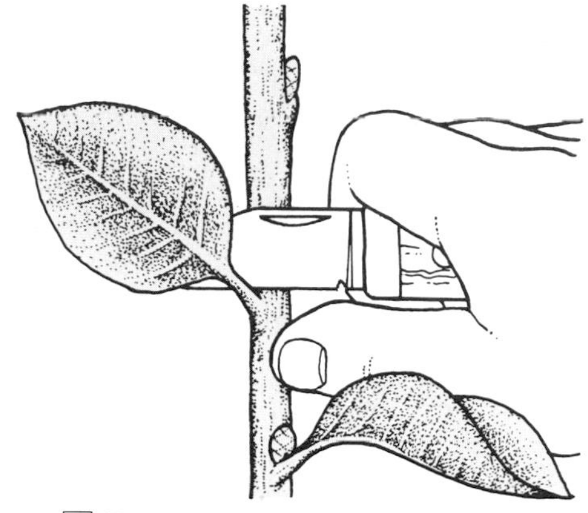

[1] Removing leaves and leaf stalks.

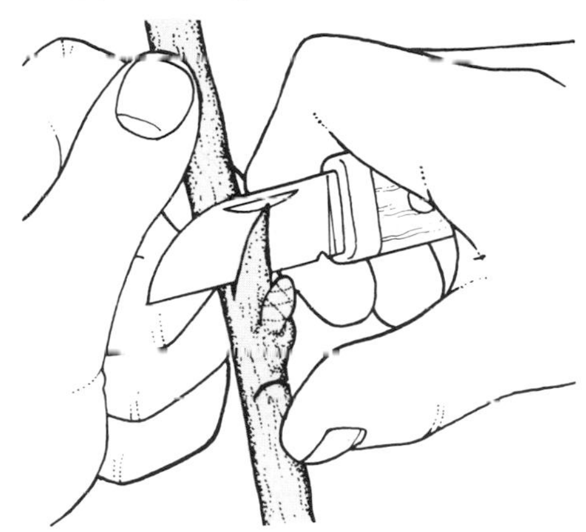

[2] Cutting out bud.

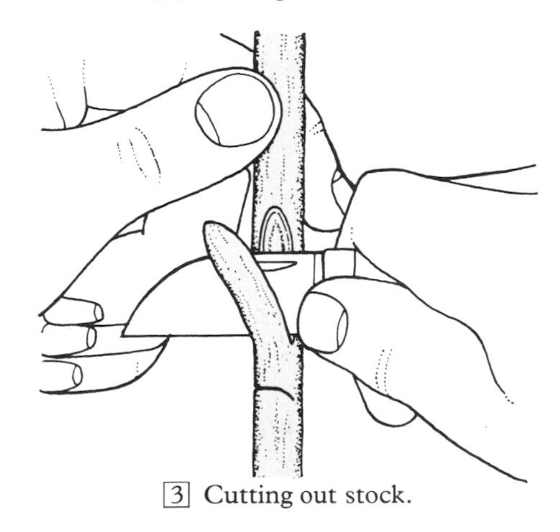

[3] Cutting out stock.

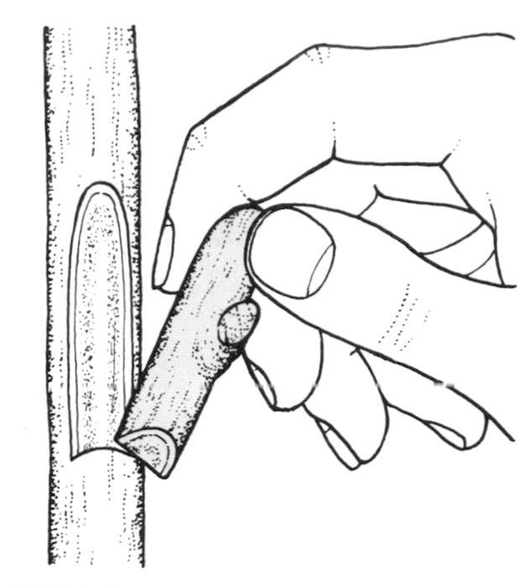

[4] Fitting bud on to rootstock.

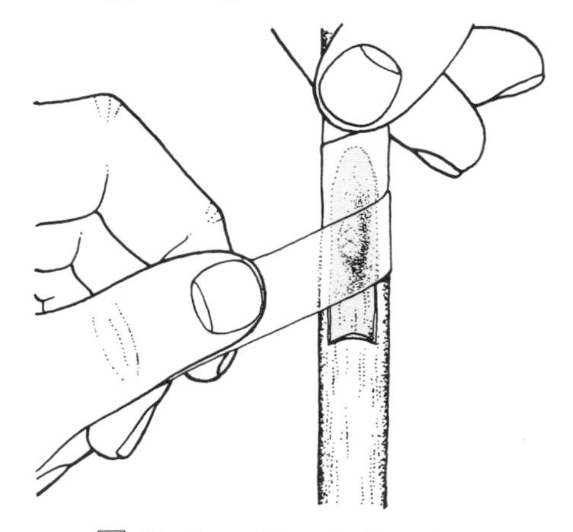

[5] Taping with polythene tape.

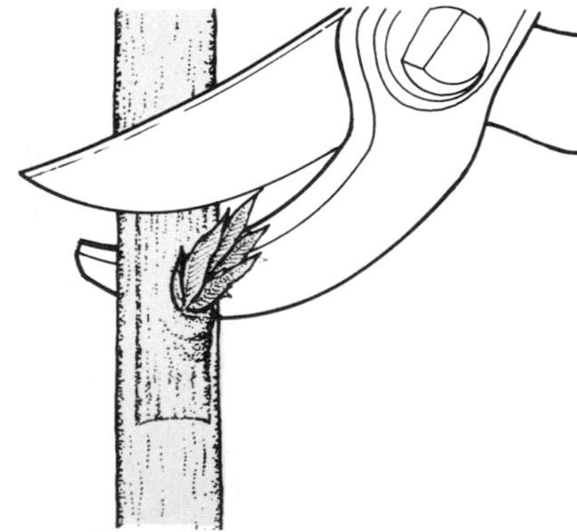

[6] Cutting back rootstock.

[4] Fit the bud on to the rootstock, pushing the wedge-shaped base down into the wedge cut on the stock. If the two edges of the cuts meet all the way round there will be good cambial contact. Even if the edges are only matched on one side, there should be sufficient contact for a take, though clearly the better the match, the better your chances.

[5] The completed chip bud should be firmly tied and sealed, and clear polythene tape is excellent for this job. Tie it below the bud, then wind the tape round the stem in an overlapping spiral. Cover the bud completely, but be sure to adjust your overlaps so that there is only a single thickness of tape over the bud itself. Tie the tape off above the top of the bud.

The bud should have united with the rootstock within about a month. As long as there is just a single layer of polythene over the bud, as it swells it will burst through the tape and grow away with no ill effects; however, it does no harm to carefully slit or remove the tape once you see the bud swelling.

[6] Once the bud is obviously growing strongly, the rootstock can be cut back to just above it. If you chip bud in early spring, you should be able to cut the rootstock back within a few weeks.

The other time you might chip bud is in late summer during dry weather, when the rind does not lift easily enough for shield budding. In this case, growth from the bud will start the following spring, and the rootstock can be headed back in late winter.

Whip and tongue grafting

Whip and tongue grafting uses a larger portion of shoot for a scion, not just a bud. The principle of cambium contact is just the same, though.

[1] The plant chosen as the rootstock usually has some special benefit for the plant that is to be propagated. Some slow growing ornamental plants need a vigorous rootstock to promote good growth: fruit trees however often require the opposite – a dwarfing rootstock which keeps leafy growth under control and encourages early cropping. Sometimes a rootstock is resistant to soil-borne diseases that would cripple the scion variety.

Select the rootstock carefully and establish it in the garden, letting it grow for a full season before grafting.

[2] In the middle of the dormant season, collect scions of the plant you want to propagate. Choose plump, healthy stems of the previous year's growth and cut them with secateurs just above a bud.

[3] Grafting is done in very early spring, just before the buds burst. The scion wood should, however, be fully dormant if possible, which is why it is collected earlier in the winter. To hold them at the right stage, the labelled scions are bundled together and their bases buried in about 6 in (15 cm) of soil in a cool, shady position. Make the bundles only as big as you can comfortably hold in one hand – if they are too big they will begin to heat up in the middle.

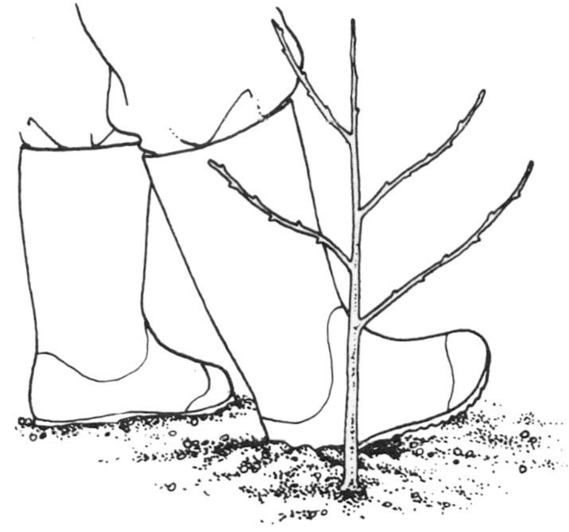

[1] Selecting suitable rootstock.

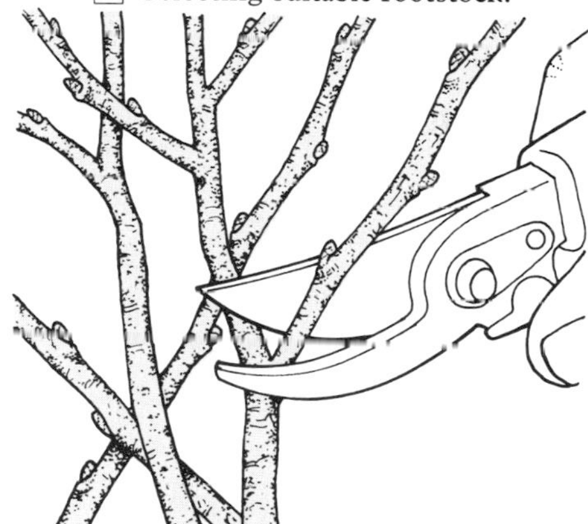

[2] Collecting scions.

[3] Heeling in scions.

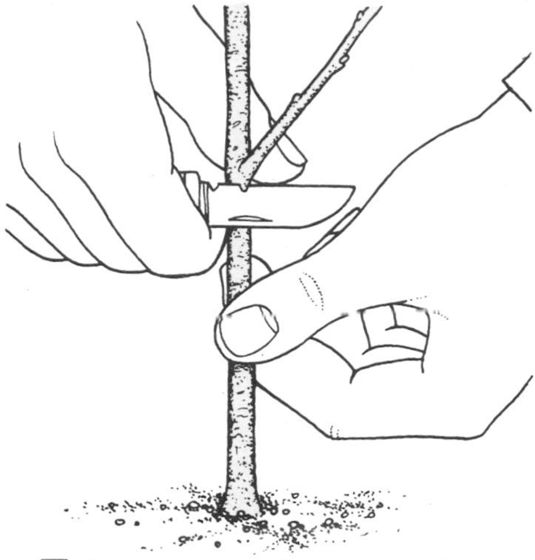

[4] Preparing stock by removing lower branches.

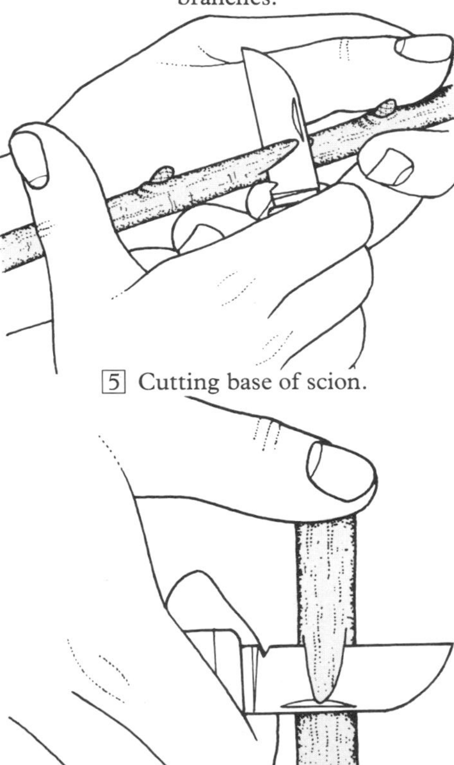

[5] Cutting base of scion.

[6] Cutting base of stock.

[4] The graft is made well clear of the soil, because in most cases the scion variety must not produce roots; if it did, this would overcome the beneficial effects of the rootstock. The grafting point is usually 6–10 in (15–25 cm) above ground level.

Prepare the rootstock in late winter or early spring by removing all the lower branches, leaving about 15 in (38 cm) of clear stem.

[5] When you are ready to graft, lift your bundle of scions and take them to the rootstocks. The lower part of each scion is used, and the base is prepared by making a sloping cut which ends just below a bud. It is important that the cut surface is perfectly flat, and it's a good idea to make several practice cuts on some spare scions first.

If you are right handed, hold the top of the scion in your left hand, and the knife, in the correct position at the base of the scion, in your right. Make the cut by drawing your two hands slowly and steadily apart in a single smooth action. You have more control if you keep your hands close in to your body, but do remember that knife is sharp – and dangerous!

[6] Using a sharp pair of secateurs, cut the top off the rootstock at approximately the right height. Then using the same smooth, single action with the knife, make a sloping cut on the stock that will exactly match that on the scion. If you are a beginner, it's a good plan to allow yourself some leeway and start higher up the stem; you will probably need to make several cuts before you get it just right. Practise making matching cuts on odd pieces of branch first, before using your precious scions and rootstocks.

[7] To make sure the scion and stock stay firmly together, a tongue is cut in each. Cut the stock first, by making a shallow, straight cut about one-third of the way down from the top of the sloping surface.

[8] A corresponding cut is made in the sloping base of the scion, but this time about one-third of the way up from the lowest point. Be careful not to cut too deeply.

[9] Cut the top off the scion so that it has three good buds. Fit the scion and rootstock together so that the two tongues interlock, and check that the two cambium layers are in contact. Look at the graft straight on from the side – if you can see daylight between the two surfaces, the cut surfaces are not perfectly flat and they will not unite.

If the stock and the scion are the same diameter, you only need to concentrate on getting your carpentry right. Often, though, the rootstock is a bit larger than the scion, and the temptation is to put the scion right in the centre of the cut on the stock, resulting, of course, in no cambium contact and no new plant. If you can't exactly match stock and scion sizes, set the scion to one side so that at least there is cambium contact at one point.

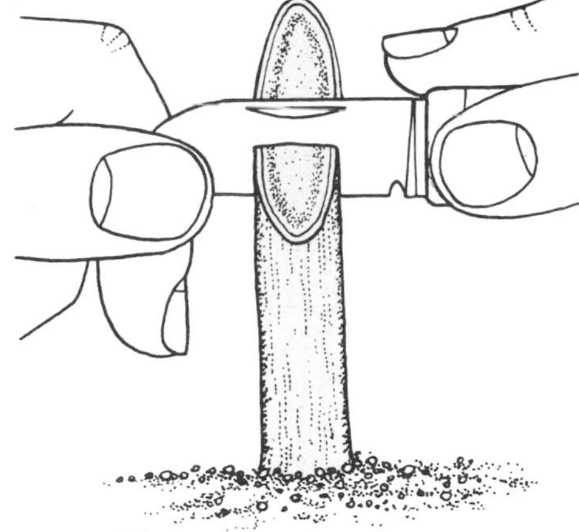

[7] Cutting tongue on stock.

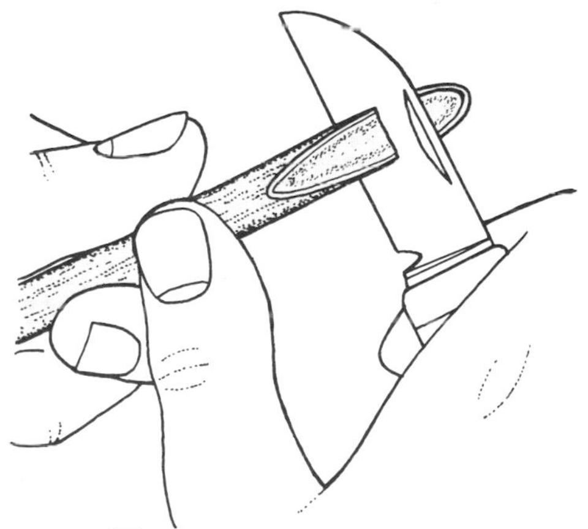

[8] Cutting tongue on scion.

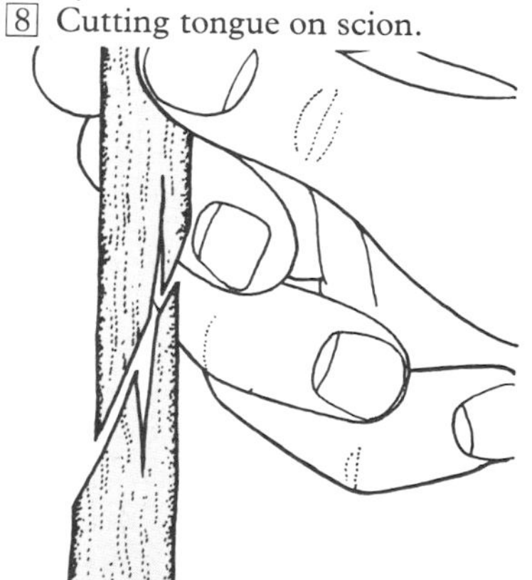

[9] Fitting same-size stock and scion together.

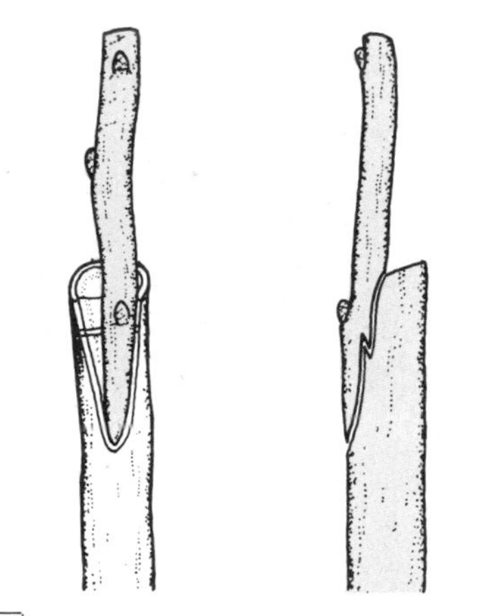

[10] Fitting small scion on to larger stock.

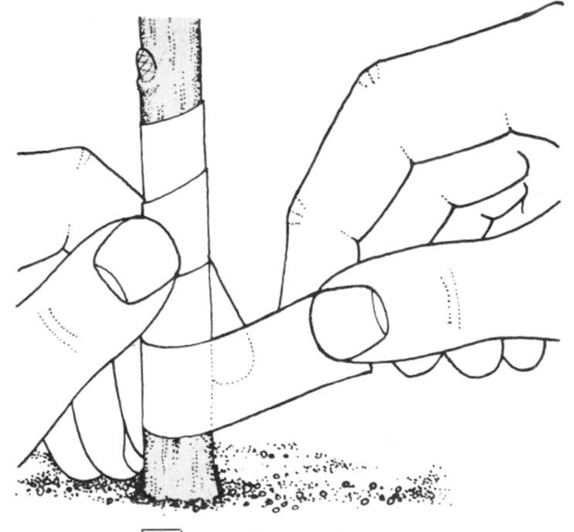

[11] Binding with tape.

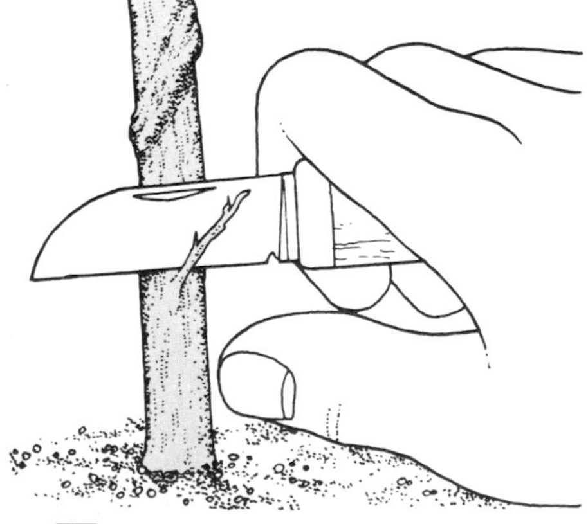

[12] Removing growths from stock.

[10] Occasionally you may need to graft a scion on to a very much larger rootstock. In this case it is easier to make the sloping cut down just one side of the stock, having cut the top of the rootstock off with secateurs as before. Set the scion to one side where necessary to get maximum possible cambium contact.

[11] Although the tongues should mean a secure join between the two plant portions, the graft will need wrapping to protect it from movement and from drying out. Polythene tape, wrapped tightly round the area, is good where the stock and scion are the same size. Where they are different sizes, and the scion is offset or on one side of the stock, raffia is easier to handle: be careful not to knock the scion out of position while you are tying it. Having bound the graft firmly with raffia, the whole of the union can be covered with grafting wax or a proprietary tree wound-sealing compound, particularly exposed cut surfaces. This protects them from drying and from disease attack. Polythene tape does not need painting over with sealant, but a dab on the cut surface of the top of the scion is a good idea.

[12] Corky callus tissue will form at the points of contact between the surfaces, and the two parts will unite. Eventually shoots will grow from the buds on the scion. Slit the tying material with a sharp knife before it constricts the swelling stem.

There may be some growth from below the graft union, too: this must be cut off with a sharp knife as soon as it is noticed. Let the rootstock shoots develop too much and they will be difficult to distinguish from the scion growths.

Spliced side graft

There are all sorts of different ways of fitting stock and scion together to get good cambium contact. Whip and tongue is one of the most straightforward, but the spliced side graft is often used for conifers, and is slightly different.

[1] Pot up a young conifer which will be suitable as a rootstock, and leave it to establish in its pot for a complete season.

[2] In late winter, collect a scion from the variety that is to be propagated, selecting a strong, healthy shoot which grew the previous year. The base of the shoot should be mature wood.

Remove the lower leaves from the scion, and make a long sloping cut 1–1½ in (2.5–4 cm) long on one side: the wood can be difficult to cut smoothly so make sure you start with a really sharp knife. On the opposite side of the stem, make a much smaller sloping cut to form a short wedge at the base.

[3] Trim off any shoots growing at the base of the rootstock to give you a length of clear stem. Starting about 2 in (5 cm) from soil level, make a shallow, sloping cut in one side of the rootstock stem, making this the same length as the long cut on the scion. Withdraw the knife.

[4] Make another short, slightly sloping, downward cut on the stock to meet the base of the long cut you have just made. You can now remove the sliver of wood with its wedge-shaped bottom, very similar to the arch-shaped cut described for chip budding earlier on.

[1] Planting conifer rootstock in pot.

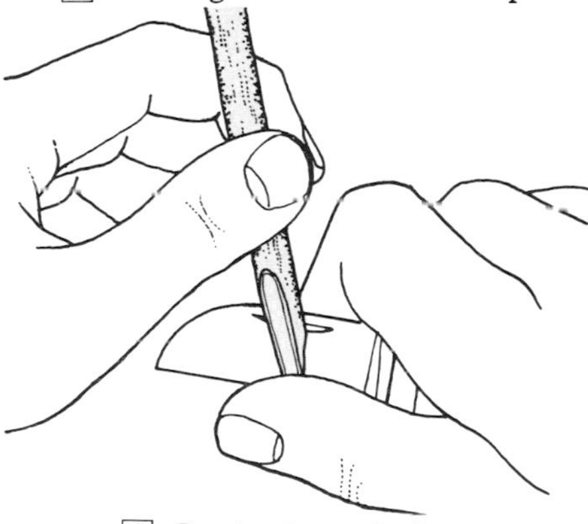

[2] Cutting base of scion.

[3] Making long cut on stock.

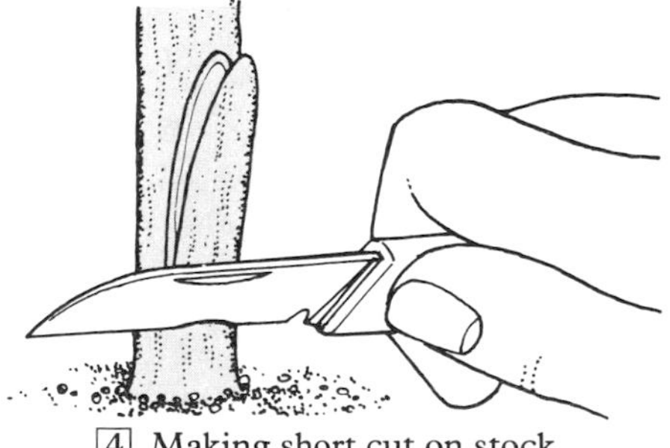

[4] Making short cut on stock.

[5] Fitting stock and scion together.

[6] Union tied with raffia.

[7] Potted graft placed in mist unit.

[5] Fit the scion and rootstock together so they are held in position by the wedges at their bases. You should be able to make the cut on the stock exactly the same width as the scion, so that the two fit together perfectly with cambium contact all round. Sometimes the scion is very narrow and this isn't possible, but as always, make sure the cambium layers match at some point.

[6] Tie the stock and scion together firmly with raffia or polythene tape. Note that with this graft, the head of the rootstock is left on for the time being, unlike the whip and tongue graft where the rootstock is cut back right from the start. Leaving the head on gives the plant less of a shock and helps speed healing of the graft.

[7] The newly grafted plant should be kept in a warm, humid atmosphere while the two parts unite. An automatic mist unit is ideal, providing the perfect conditions. If you haven't a mist unit, keep the pot in a warm greenhouse under a plastic cover, and spray it over with plain water from a hand sprayer.

The plants should have united within about two months, when the raffia or tape can be removed. Take the plant out of the very humid conditions and cut the rootstock head back by half. After a further two weeks, cut the rest of the rootstock head off just above the graft, and harden the plant off ready for planting outside.

Topworking established fruit trees

An established fruit tree is always a bonus in any garden, but it can turn out to be a bit of a disappointment. Often it's a poor variety, producing meagre crops of flavourless apples that won't keep for more than a fortnight and usually end up on the compost heap.

All is not lost however. With a bit of skilled grafting you can magically change your tree to any variety you like – even two or three different varieties at once!

[1] Regrafting an entire apple tree is a big job, but a satisfying one. The simplest method is to remove virtually all the branches, setting scions of the new variety in the stumps of the four or five main branches. This is known as topworking.

In midwinter, look at the tree and see how many main branches there are arising from the crown. Imagine these to be cut off about 2–3 ft (0.5–1 m) from the crown, and count how many cut ends you will then be left with. Allowing four or five scions for each branch (you won't need all these, but it's best to be on the safe side), estimate the total number of scions you will need, and collect them from your own trees or beg them from friends and neighbours. Bundle them up and heel them in to a cool spot as described in whip and tongue grafting.

In early spring, around February, saw the branches off the tree to within about 3 ft (1 m) of the crown, leaving a well-spaced framework. The diameter of the sawn-off branches should not be more than about 4 in (10 cm).

[2] There's a variety of ways in which you can insert the scions into the branches. Cleft grafts are usually made with a special clefting tool, but a well-sharpened billhook can be used instead. The blade is tapped firmly into the stump with a mallet, cutting across the whole surface.

[3] Once made, the cleft is kept open by inserting a wooden wedge into the centre, the wedge being made by

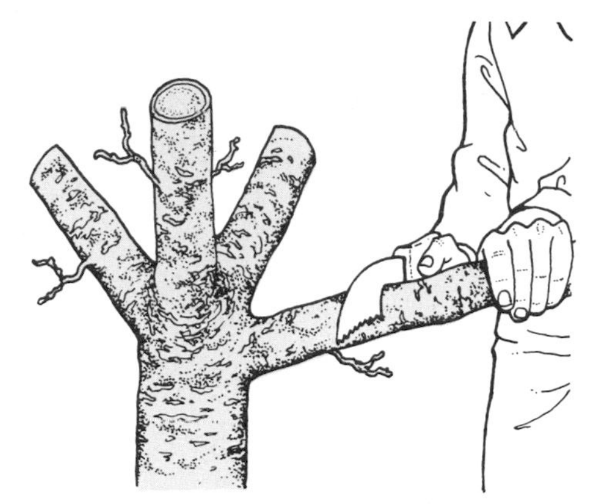

[1] Sawing branches off apple tree.

[2] Making cleft in branch stub.

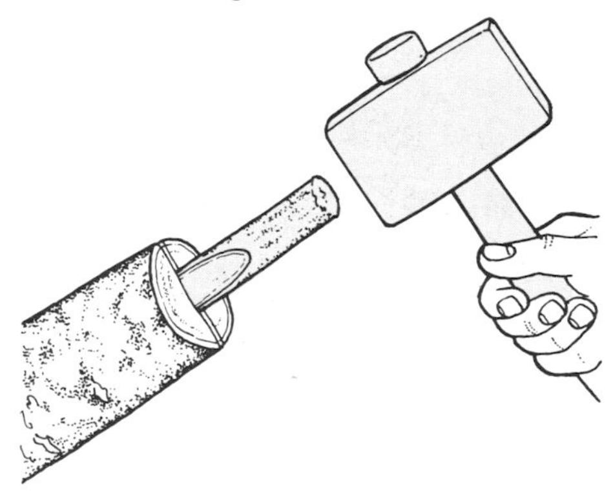

[3] Inserting wooden wedge.

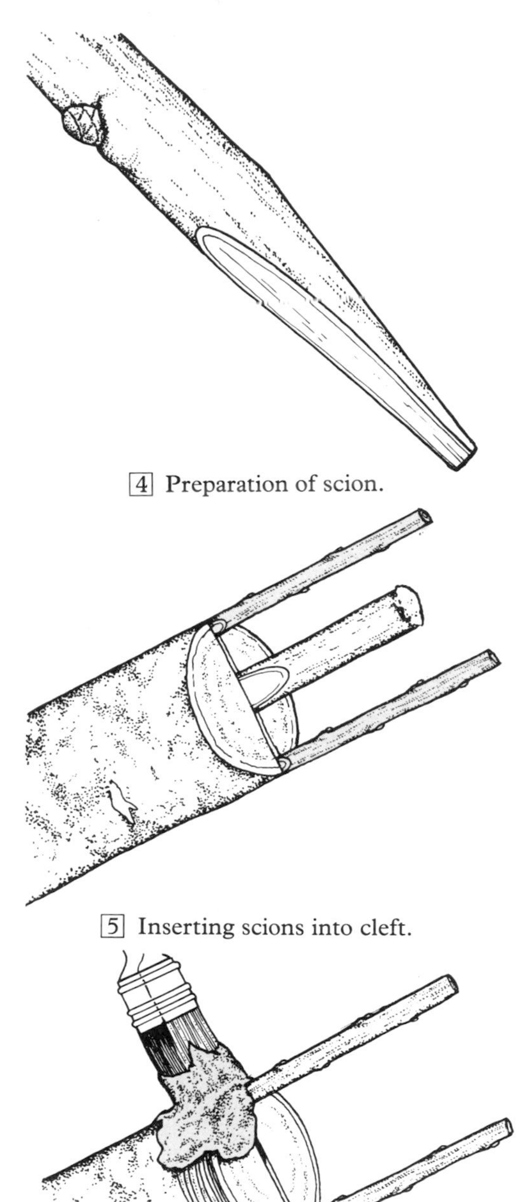

[4] Preparation of scion.

[5] Inserting scions into cleft.

[6] Applying wax.

trimming both sides of a suitably sized stick.

[4] Two scions are used in each cleft graft. They are prepared so that they are three or four buds long, with both sides of the base trimmed to give a long, tapering wedge. As with whip and tongue grafting, it's important to make these cuts smooth and level.

[5] Insert the scions into each side of the cleft, so that the cambium layer of the branch stub matches the outer cambium layer of the scion. It's very important to get the position of the scions right. Push them firmly down into the cleft, trimming off their bases if necessary to get a snug fit. Remove the wedge (being careful not to knock the scions while you're doing it) and the cleft will grip the scions firmly in place.

[6] The completed cleft grafts should be sealed with grafting wax or a sealing compound. Large clefts may be filled in first with clay or Plasticine to avoid using too much wax. Some hours after applying the sealant, check that it is, in fact, still covering all the surfaces, and give another coat if necessary.

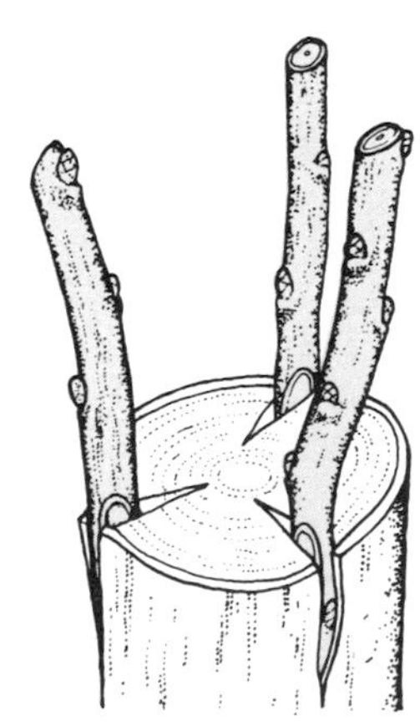

❀ Scions inserted in oblique clefts.

❀ With a very large branch stub, it's sometimes difficult to make the cleft right across the top, particularly if you don't have a proper grafting tool. In this case make three short oblique clefts round the edge of the branch stub, inserting the scions in the same way as before. Although you should keep the diameter under 4 in (10 cm) wherever possible, sometimes larger stubs are unavoidable.

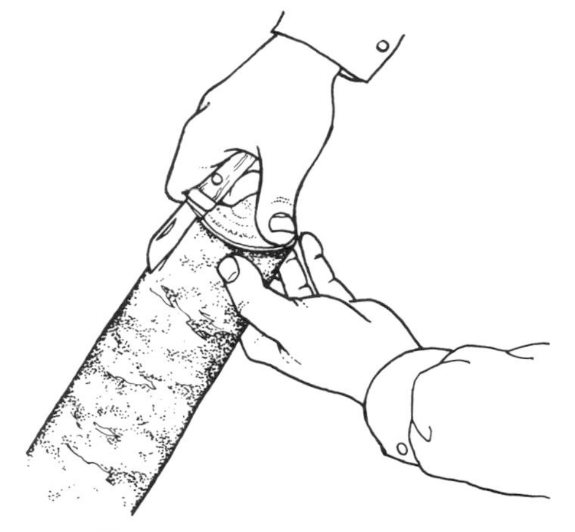

[1] Cutting rind for rind grafts.

Rind grafting

[1] Another method that can be used to topwork trees a little later in the season is rind grafting. This can only be done when the sap is rising and the rind (or bark) lifts away easily from the wood. April is usually the best month.

Make a 2 in (5 cm) downward cut in the rind of the sawn-off branch, and lift the rind away slightly to expose the cambium (like shield budding).

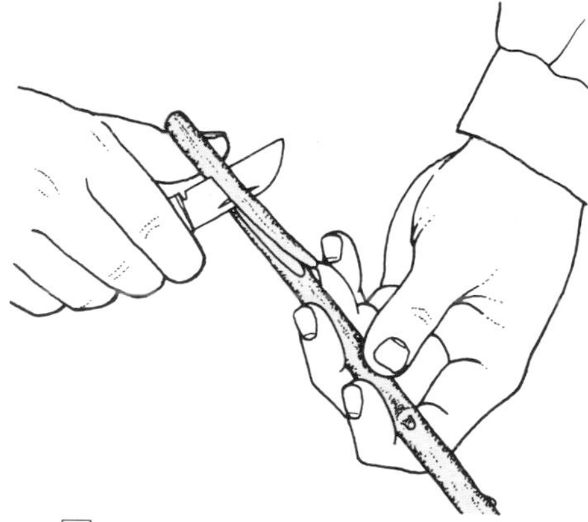

[2] Preparing scion for rind graft.

[2] Prepare the scions by cutting off the tips, leaving three or four buds, and making a long, straight, sloping cut at the base. Two to four scions can be used on each branch, depending on the diameter of the branch stub.

[3] Insert the scions behind the lifted rind so that their cut surfaces face the centre of the branch – like inserting the bud into the T-cut when shield budding. The rind is thicker and not so easy to work with as the rind on a rose, but even so, the scion should be fairly easy to insert. You may need to recut the base of the scion to get a good fit, but don't fiddle about with it too much. If you do need to recut, always make a completely new cut; don't try to alter the old one or you'll end up with a wavy surface which won't get you anywhere.

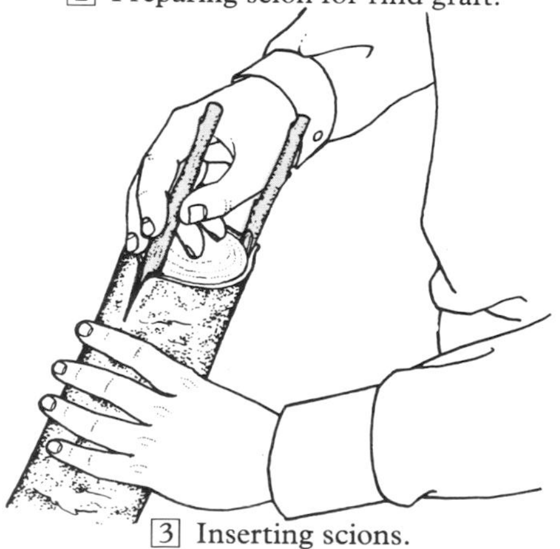

[3] Inserting scions.

[4] Binding with raffia.

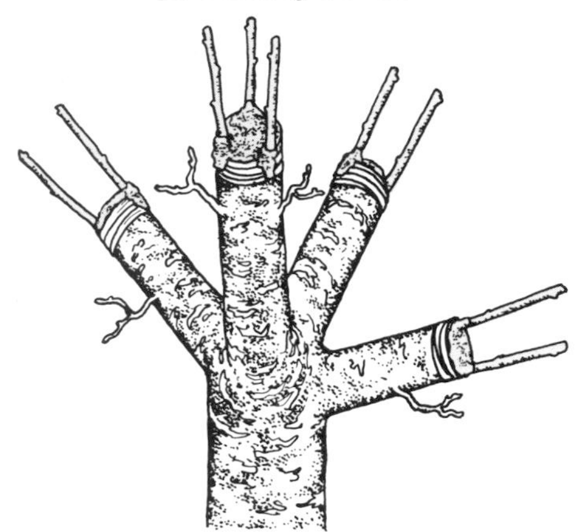

☼ Complete topworked tree.

[4] Rind grafts need tying to hold them in place, unlike cleft grafts, which are gripped naturally by the branch. Use strong, thick strands of raffia, or soft string such as fillis. Two gimp pins can be hammered through the centre of the scion into the branch; this rather drastic measure doesn't appear to do any harm to the graft, but shouldn't be necessary with good craftsmanship!

Once the rind grafts are tied in place they should be sealed with wax or pruning paint. Cover all the cut surfaces with the sealant, including the top of the scion and the split bark where the scions are inserted.

☼ The completed tree will certainly look unpleasantly mutilated to start with! The small branches below the grafts are left on the tree as sap drawers: these and suckers from the stock will grow vigorously. They can be thinned out lightly if they threaten to take the tree over, but otherwise they are left alone for at least a season. In the first winter after grafting they can be very much reduced in quantity, and by the second winter they can be removed altogether.

After this, sucker growths from the original tree must be removed, and the growth from the scions pruned to form a good framework. It's important not to neglect a topworked tree, or all your hard work will be wasted. After three or four years, the renewed tree should be cropping.

Stub grafting

Many lone fruit trees don't crop well because they need a different variety to pollinate their blossoms. If you haven't the room to plant another tree, you can graft some branches of a pollinator variety into your existing tree.

[1] Choose a variety of the same fruit (apple, pear etc.) that is suitable for pollinating your tree. You can either have another fruiting variety, or a straight pollinator such as a type of crab apple.

Cut the scions to three or four buds, with their bases trimmed to an uneven wedge shape.

[2] Choose a branch of your fruit tree to carry the pollinators. The branch should have several strong laterals (side shoots) of less than 1 in (2.5 cm) in diameter into which the scions can be inserted.

Make a cut in the upper side of the lateral, beginning about ½ in (1 cm) away from where it joins the branch, and sloping towards the branch. Don't cut too deeply; about one-third of the way through the lateral is enough, and the cut certainly shouldn't go any deeper than half way.

[3] Grasp the end of the lateral and bend it down sufficiently to open up the cut (but not enough to snap it off!). Insert the scion so that the cambium layers match on at least one side – if scion and lateral are the same size there will be a good match all round. Once the scion is snugly inserted, let go of the lateral and the springiness of the branch will hold the graft firmly in place. This is known as a stub graft.

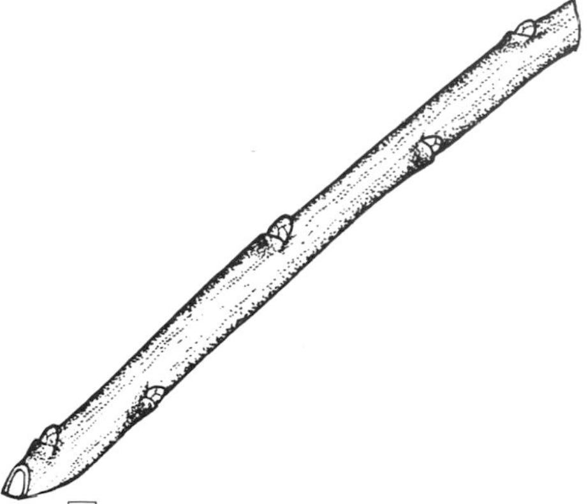

[1] Cut scion with trimmed base.

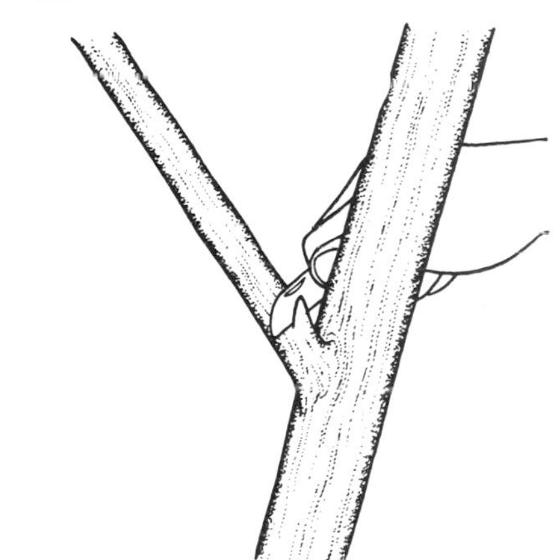

[2] Making cut in upper side of lateral.

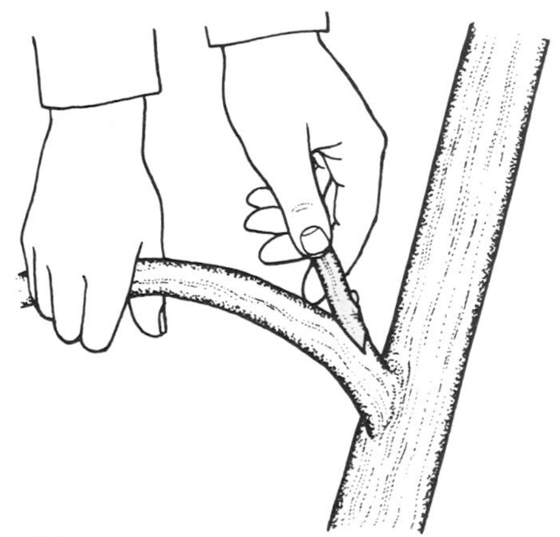

[3] Bending lateral and inserting scion.

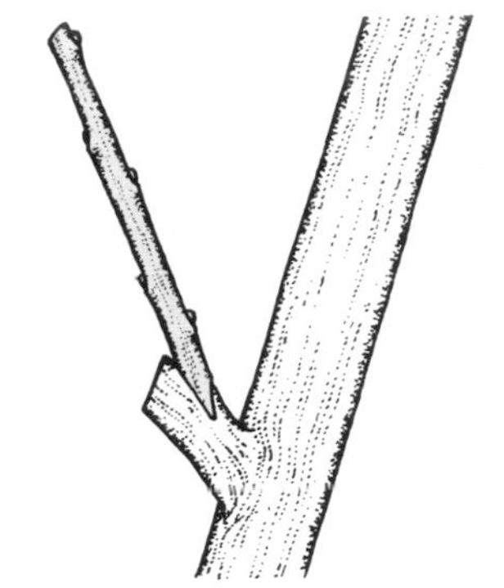

[4] Lateral severed just beyond junction.

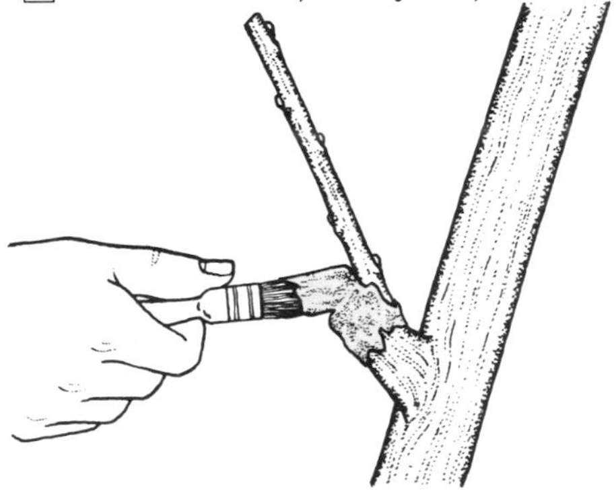

[5] Sealing the graft and the cut lateral.

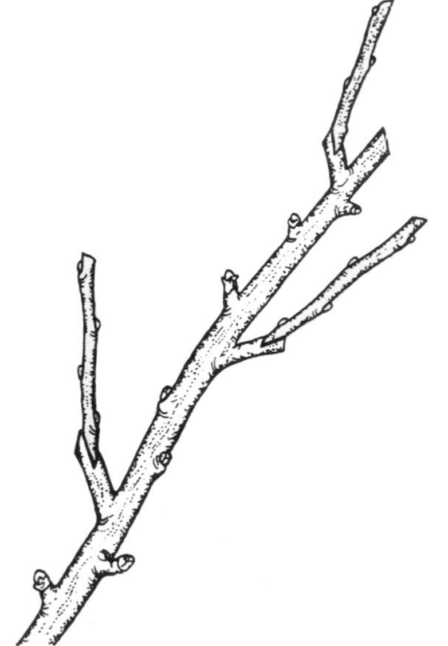

[6] Main branch cut after last stub graft.

[4] Once the graft is completed, cut the end of the lateral off just in front of it. A pair of sharp secateurs is the easiest tool to use, but you will need to be careful not to snip the newly inserted scion, or at least dislodge it.

[5] Paint the grafts and the cut end of the lateral with a wound sealant to prevent drying out and disease attack. Because the graft is held firmly in place by the natural spring of the lateral it will need no further fixing.

[6] Continue working up the branch until all suitable laterals have been grafted with the pollinator variety. Usually one large branch per tree is sufficient. Cut the top off the branch just above the last stub graft, and treat this cut, too, with a wound sealant.

Label the pollinator branch clearly at several points with long-lasting tags. It may be obvious which it is when it has just been grafted, but after a few years it will merge into the rest of the tree, and you could make a disastrous mistake at pruning time!

'Family trees', available at quite a high price from garden centres, consist of three different varieties on the same tree. By a combination of the topworking and stub grafts described you can make your own family tree, overcoming pollination problems as well as giving yourself a better choice of fruit.

Saddle grafting

Saddle grafting can be used for almost any subject where the scion and stock are of the same size, but because it is a tricky graft it isn't a common method. It does, however, give good results with rhododendrons.

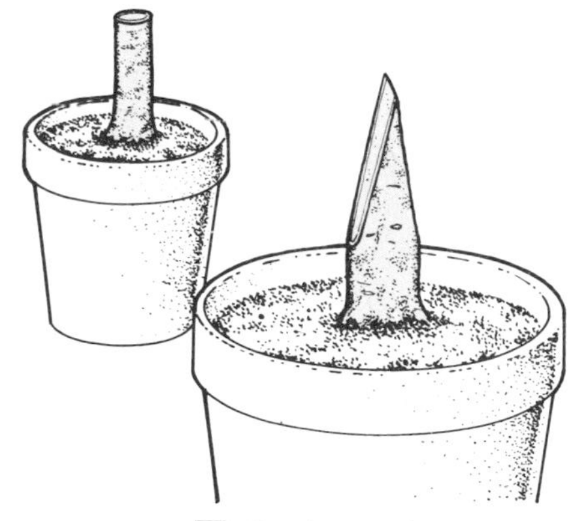

[1] Cutting stock.

[1] *Rhododendron ponticum*, the wild rhododendron, is used as the rootstock for various rhododendron varieties: grafting takes place in early spring.

Cut the stock to a wedge shape with long, even, sloping sides. Make the cuts as low as you can on the stock.

[2] Cutting scion.

[2] The scion should be as near as possible to the size of the rootstock. Carefully cut the base of the scion to a wedge shape to exactly match the cuts made on the stock. This is not all that easy to do – sharpen your knife well before trying it!

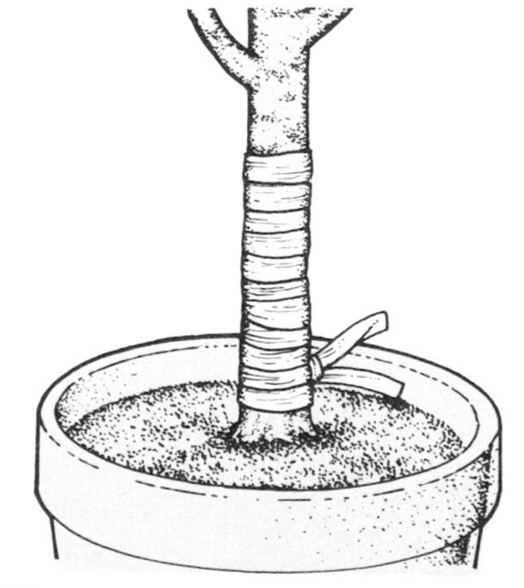

[3] Stock and scion bound with raffia.

[3] Fit the scion over the stock like a saddle, ensuring the cambium layers match. Bind the union firmly with raffia to hold it in position.

Put the grafted plant in a warm, humid atmosphere. A mist unit is ideal, but a closed propagating case will do quite well. Callus tissue will form and the stock and scion will unite, though the process may be a slow one.

Once the union has been made the plant should be gradually hardened off. Grow it on in its pot until it is large enough for planting in its permanent position in the garden; this may be two or three years.

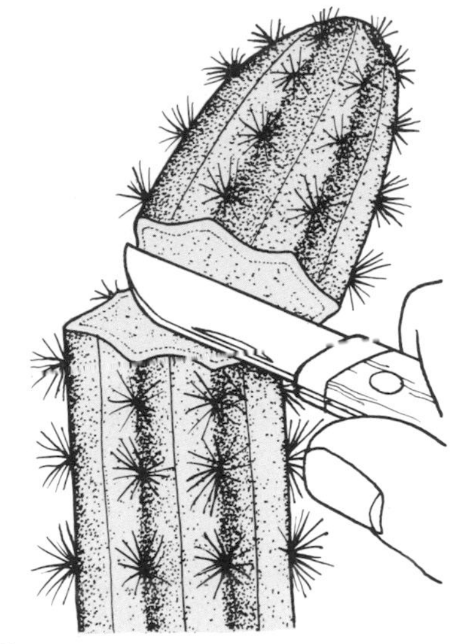

[1] Cutting top off cactus stock.

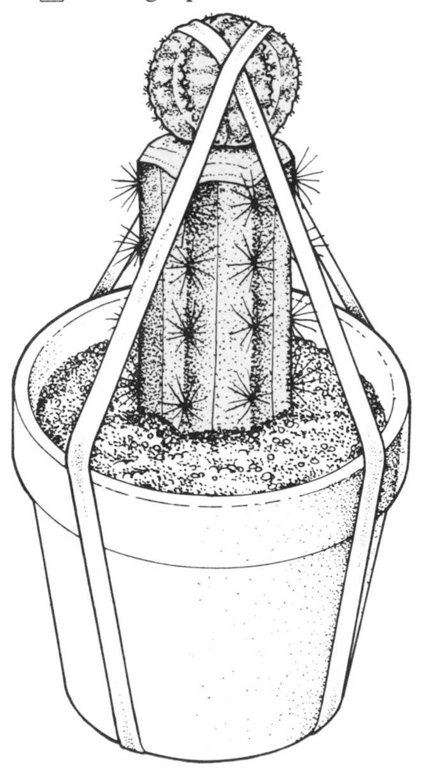

[2] Scion and stock sealed with rubber bands.

Grafting cacti

Cacti are grafted not so much to perpetuate the species or increase your stock as to obtain interesting and unusual plants by joining two different types. Slow-growing cacti can be grafted on to a faster growing stock, too, to speed them up. It is quite a simple procedure, though the formation of the union sometimes takes a long time, and can sometimes fail altogether for no apparent reason.

[1] Early summer is the best time to try grafting cacti, when the whole of the growing season is ahead. Choose a tall growing, green variety as the stock – *Trichocereus* or *Hylocereus* are suitable types. Slice off the top 1–2 in (2.5–5 cm) with a sharp knife so that it leaves a flat surface. Trim the edges of the cut lightly to round them off slightly.

[2] Choose a scion that is of a similar size to the stock. For the best effect pick a differently shaped cactus – a round scion on a cylindrical stock is the most popular combination. Cut a thin sliver off the base of the scion so that it is quite flat.

Look at the cut surfaces and you will see the dark ring of tissue that is equivalent to the cambium layer: match this in at least one place. Hold the scion in position with elastic bands that will hold the graft firmly but not too tightly.

Leave the plant undisturbed for several weeks, until it is clear that the graft has taken.

Grafting tomatoes

Tomatoes are often grown in the same greenhouse soil for years, which encourages a build up of soil-borne pests and diseases. When these are present, normal varieties of tomato will suffer; the crop will be severely reduced or the plants may fail completely. Soil sterilization or a complete soil change is often impractical, but the problem can be overcome by grafting fruiting varieties on to a pest and disease resistant rootstock such as KNVF.

[1] Establish both the fruiting and rootstock varieties in pots, planting them at the edge. Make sure each is clearly labelled! They are ready to graft when the seedlings have about four true leaves. Remove the seed leaves if these are still present. Take a small slice off the sides of each of the stems, making the cuts exactly the same size and height.

[2] Place the two cut surfaces of the stems together and tape them in place with a single thickness of Sellotape. If necessary, first remove the plants from their separate pots and replant them together in a single container.

[3] The plants should have united within about three weeks. The tape can be carefully cut away, and the top of the rootstock, and the roots of the fruiting variety, cut off with a razor blade. Make sure the graft union remains well above the soil when planted.

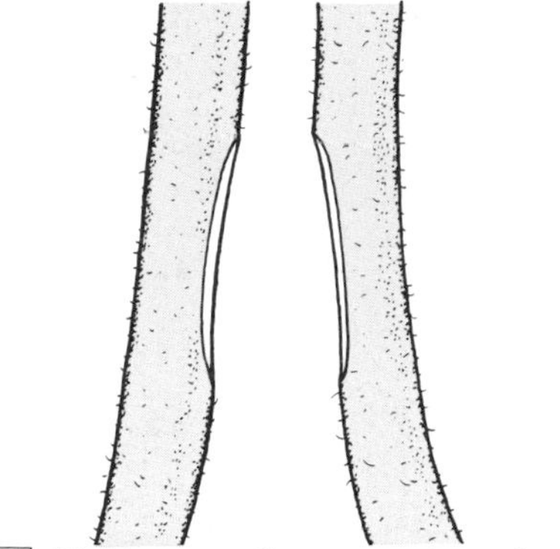

[1] Making cuts for approach graft on tomatoes.

[2] Taping with single thickness of Sellotape.

[3] Top of stock and root of scion are removed.

What might go wrong

The grafted scion fails to grow
•Poor carpentry could mean that the two surfaces fail to fit together smoothly. •The scion has been badly positioned on the stock. •The cambium has not been sufficiently exposed on the stock or scion. •The scion has been knocked out of place because it was not securely fixed. •The rootstock and scion are not compatible varieties. •You are trying to graft at the wrong time of year.

The rind will not lift from the wood for shield budding or rind grafting
•The sap must be rising before either of these methods can be used – there's no point in trying them in the dormant season. •In long, dry spells, keep the stock plants well watered all through the spring and summer before grafting.

The scion starts to grow, then dies or breaks at the union
•Some scions and rootstocks show what is known as 'delayed incompatibility'; they only decide they can't get on together after growth has begun. •The scion wood may be too advanced at grafting time. •Drying winds or hot sun could be too much for the young scion growth.

The beneficial effects of the rootstock don't materialize
•Dwarfing and pest and disease-free rootstocks won't do the slightest good if the scion is allowed to root, too. Unless scion rooting is to be encouraged (with rhododendrons and roses, for example), keep the graft union well above soil level when planting. Don't allow soil to heap up round the plant's stem.

Plants suitable for grafting

The most important group of plants to be budded or grafted is probably fruit trees, where the rootstocks serve a particular purpose – for example:
Apple rootstocks M27 (very dwarfing), M9 (dwarfing), M26, MM106 (semi dwarfing), MM111 (vigorous).
Pear rootstocks Quince A (semi vigorous), Quince C (dwarfing).
Plum rootstocks Pixy (very dwarfing), St Julien A (dwarfing), Myrobalan (semi vigorous), Brompton (semi vigorous).
Peach, nectarine and apricot rootstock St Julien A (dwarfing), Brompton (semi vigorous).
Cherry rootstocks Colt (dwarfing), Mazzard (vigorous), F12/1 (semi vigorous).

A wide range of ornamental plants can also be grafted: generally this method is used where a named variety, having double flowers, variegated foliage, or some other variation on the 'type' species is to be increased. These variations will not come true from seed and are often difficult to grow from cuttings: they may also be very weak growing, and need a stronger rootstock to give them extra vigour. The rootstocks used are usually seedlings of the type species, e.g. *Gypsophila paniculata* for *Gypsophila paniculata* 'Bristol Fairy'. Named varieties of the following plants are often grafted: *Acer*, *Clematis*, *Cornus*, *Crataegus*, *Gypsophila*, *Hamamelis*, *Ilex*, *Magnolia*, *Malus*, *Prunus*, *Rhododendron*, *Rosa*, *Syringa*, *Viburnum*, *Wisteria*.

The type of graft you use is mainly a matter of personal choice. Fruit trees are usually grafted by whip and tongue in early spring and any failures are shield budded later in the season. Roses are virtually always budded. Spliced side grafts are particularly suitable for conifers and saddle grafts for rhododendrons.

5
DIVISION

Equipment

This is probably the simplest and most basic way to increase any plant, and requires the very minimum of tools and special equipment.

A border fork (*a*) is necessary for digging up the plants before dividing them. The classic way to split them into smaller pieces is to use two border forks inserted into the clump back to back – if you have two forks, that is. You may well be able to lay your hands on an old fork which is beyond digging but is quite serviceable for the occasional division: or you might find it useful to buy one border fork (often known as a ladies' fork) and one larger digging fork (*b*). A couple of handforks (*c*) are a cheaper proposition and are quite adequate for the division of smaller clumps.

Some plants are particularly tough and resist all attempts to persuade them gently into smaller pieces. They can be cut carefully into sections with a sharp knife (*d*).

Large plants in heavy soil may be more successfully dug up with a spade (*e*) than a fork, but a trowel (*f*) is fine for many herbaceous and alpine subjects.

Much division takes place outside, with the new sections being planted direct back into the garden soil. A range of the usual pots and seed trays is useful for growing on the very small sections that need a little extra care and molly-coddling: you can use garden soil for outdoor subjects, or loam-based potting compost such as John Innes No 1 or 2.

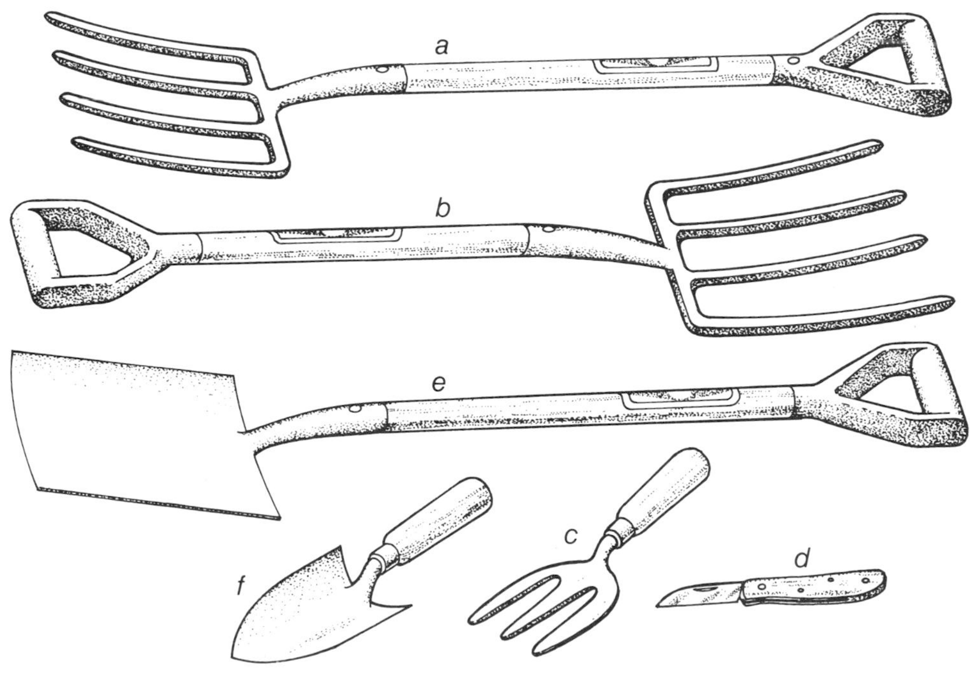

Division equipment: for details, see text above.

[1] Digging up border plant.

[2] Dividing with two forks.

[3] Pulling smaller clumps apart by hand.

Division blueprint

Herbaceous border plants grow quickly and soon make thick clumps. After a couple of years, the more vigorous plants are quite likely to start to deteriorate as the shoots become overcrowded and the centres of the clumps get woody. Dividing the crowns into smaller pieces will rejuvenate the plants as well as increase your stock.

[1] Perennial border plants can be divided as they are dying down in autumn, or in spring, just as they are starting into growth. Spring division generally gives the best results, when the weather is improving and the whole of the growing season is just ahead. Wait until the plants have sturdy young shoots: March is generally the best month.

Lift the plant carefully with a garden fork, retaining a good soil ball round its roots. Once it's out of the ground, remove most of the soil by shaking and carefully crumbling it away with your hands – this way you avoid doing too much damage to the fibrous root system.

[2] Lay the plant on a clear piece of soil. If it is a large clump, insert two forks through the centre so they are touching, back to back. Make sure they are pushed right through to the base of the clump. Pull the handles of the forks apart: this will divide the clump into two pieces. A very large specimen can be split into further pieces by the same method. Two handforks can be used in the same way on plants which need slightly less force.

[3] Smaller mats of plants can be tugged apart by hand, teasing the roots gently and being careful not to snap young shoots. Many species are very

easy to pull to pieces; others need a little more patient work before they will disentangle themselves without damage.

[4] Some plants are really tough, and won't pull apart no matter how hard you try. These need to be cut into portions with a knife or spade. Make sure each section has several strong buds or shoots and plenty of fibrous root.

[5] The centres of herbaceous plants are usually hard and woody, but the growths round the edge of the clump should be strong, healthy and vigorous. These are the portions which should be kept for replanting. The woody and worn-out centres can be discarded.

The best size for each division is about 4 in (10 cm) across the crown. When replanted this will give a good flower display the same season. If you have something particularly precious that you want to increase as fast as possible it can be split into smaller pieces; as long as each piece has one shoot and some healthy roots it will survive. Very small portions are best potted up into sterile, loam-based compost to give them a good start.

[6] Replant the divisions straight away; don't leave the roots exposed to the drying air for any longer than is necessary. Quickly dig over the spot the plant was lifted from, removing weed roots and incorporating some bonemeal before replanting there.

[4] Cutting clumps with knife.

[5] Old, woody centres with fresh growths at edge.

[6] Replanting new portions.

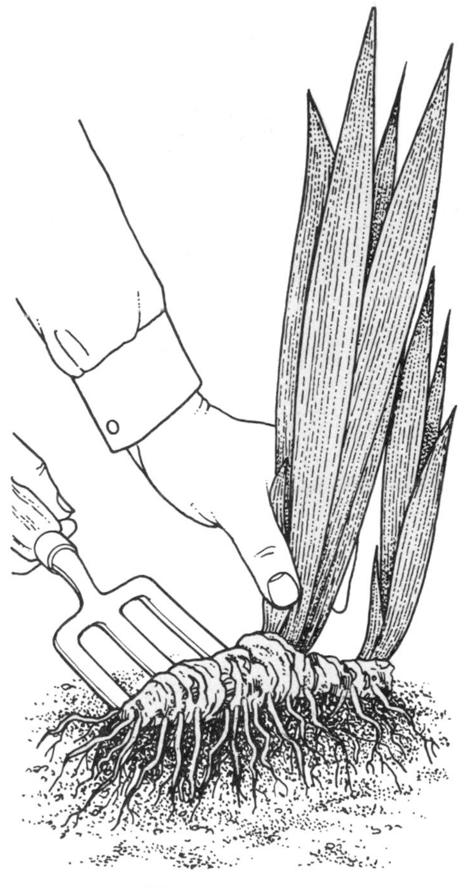

[1] Lifting irises.

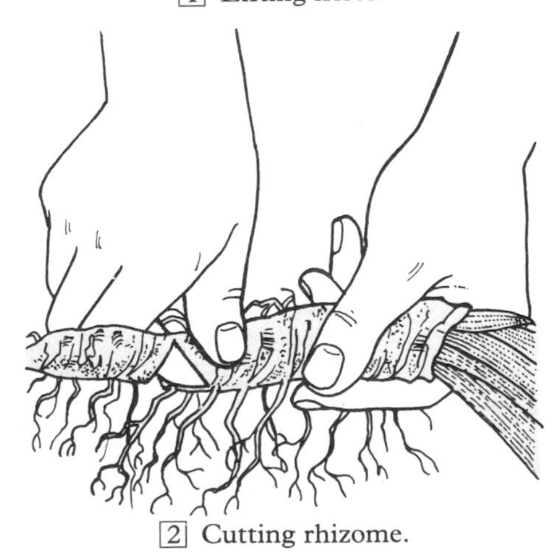

[2] Cutting rhizome.

Rhizomes

[1] Rhizomes are underground stems which are swollen and fleshy, and are often mistaken for oddly shaped bulbs or corms. They grow horizontally, producing roots from their lower surface and leaves at intervals from the tops, and they can travel a surprising distance in a garden.

Rhizomatous irises in particular form a large circle with tufts of leaves round the edge and a large dead patch in the centre, which usually becomes infested with weeds and is difficult to keep tidy. The plants should be lifted and divided every three or four years to keep them in good condition.

Lift them in summer, once they have finished flowering. Established clumps are very tough, and will need quite a bit of digging out with a sturdy border fork.

[2] With a sharp knife, cut off and retain all the healthy, vigorous rhizomes with their fans of leaves from round the edges, and discard the hard, woody portions in the centre. Sort through the rest and reject any shrivelled or damaged rhizomes, or any that have had their roots torn away.

Slice straight down through each portion of rhizome about 3 in (8 cm) in front of the fan of leaves – you will need a sharp knife as they can be quite hard. When every section is prepared it should consist of a fan of leaves and a healthy, plump piece of rhizome with plenty of roots.

[3] Having drastically reduced the amount of stem and root, you should also reduce the amount of leaf to balance the plant up. Iris rhizomes should be planted very shallowly, and they would be topheavy and unstable with their full complement of leaves. A large area of leaf surface will also lose water more rapidly than the roots can take it up. Use a sharp knife to cut the leaves down to about 4 in (10 cm) tall, following the natural fan shape.

[4] Once the sections have been prepared, they should be replanted without too much delay. Two or three pieces will replace the established clump you have just lifted; the rest will give an ample supply of new plants. They can be set out directly in their flowering positions.

[5] Dig over the area where the divisions are to be replanted, adding a little general fertiliser. Irises like a well-drained spot – if the soil is on the heavy side, add peat and sharp sand to lighten it.

Dig a shallow hole with a trowel, and press the prepared portions into it. Cover the roots with soil, but leave the top of the rhizome just showing. Firm the soil around the portions thoroughly, and water them in well. As irises are divided in summer, when the weather is usually hot and dry, you will need to make sure the soil does not dry out during the next few weeks while the plants are getting established.

Because they are divided immediately after flowering, there is plenty of time for the new plants to settle in and reach flowering size themselves by the following summer. They can be split in spring, too, but you will sacrifice that summer's blooms.

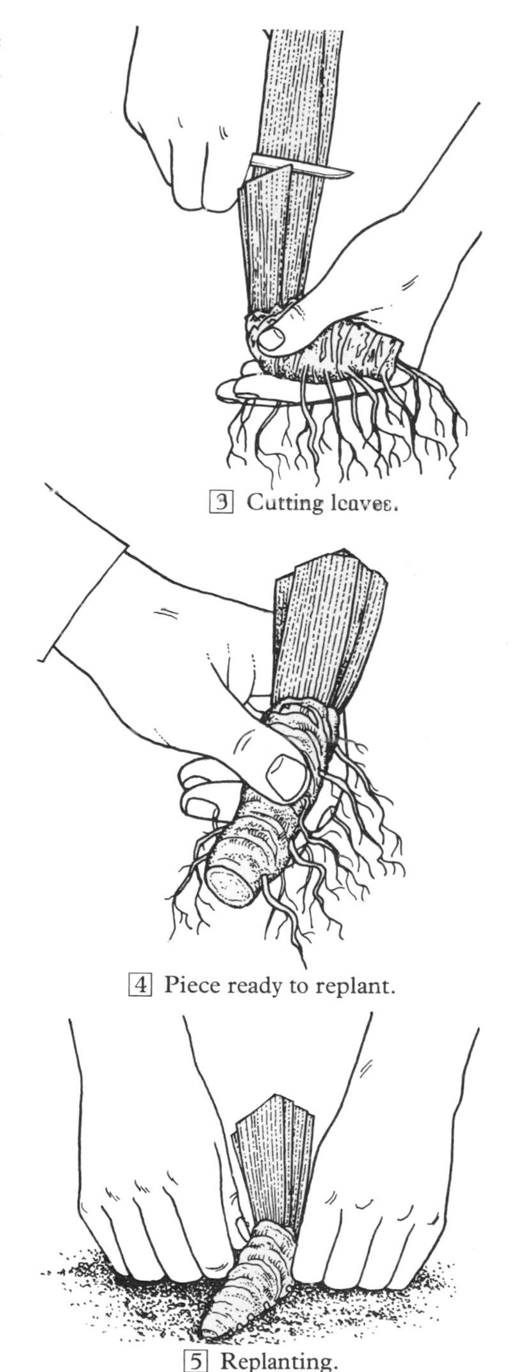

[3] Cutting leaves.

[4] Piece ready to replant.

[5] Replanting.

Tubers

Tubers are storage organs which contain food reserves to start the plant into growth each year. They are usually specialized stems which develop below ground, and have shoot buds on their surfaces. They are very similar to corms – in fact cyclamen are most commonly referred to as corms, though they are, in fact, tubers.

The tubers of some plants, such as dahlias, are not modified stems but swollen roots, and they are more correctly known as root tubers. The shoot buds on these are all found on the crown of the plant, not on the surface of the root tuber itself.

[1] Tuberous rooted begonias are popular house and summer bedding plants which can be kept for a number of years. The tubers increase in size quite considerably, and after a while the flower display does not improve, as you might expect, but becomes poorer; the crowded shoots are also more prone to disease.

Remove the tubers from their winter storage in February or early March. They have a dished shape; the side with the depression is where the shoots will appear, so this side must be kept uppermost. Press the tubers rounded side down into a tray of moist peat and keep them in a warm greenhouse. Because the centres are dished they will collect water, so don't splash the watering can around too much. Within a short while, clusters of plump buds will appear on the tuber's surface.

[2] Lift the tuber from its tray of peat and decide how many pieces it can be cut into. In theory, it can be divided into as many portions as there are buds; in practice it pays not to be too greedy. Very small pieces are more difficult to look after, and are not always successful. An average size tuber can be cut into two with good results. Use a sharp knife to divide the tuber, making a clean cut and not damaging any of the buds.

When dividing tuberous roots, each portion must include a piece of stem with a bud. The normal practice is to divide the old stem and tubers straight down the middle while the root tuber is still dormant, but if you box it up in moist peat like a begonia, shoots will soon grow and you can see exactly where to cut.

[3] The wounded edges will be prone to fungus disease attack, and should be protected with a coating of fungicide powder. This can be applied from a puffer pack, or the cut edges can be dipped in a saucer of the powder. Leave the dusted tubers out in the air for a few hours to dry off.

[4] Replant the sections in suitably sized pots of peat-based potting compost. Settle each piece deeply enough for the cut sides to be completely covered. To avoid the danger of rotting, don't cover the top of the tuber (where the shoots are) with compost. Water the plant carefully round the edge of the pot, or from below.

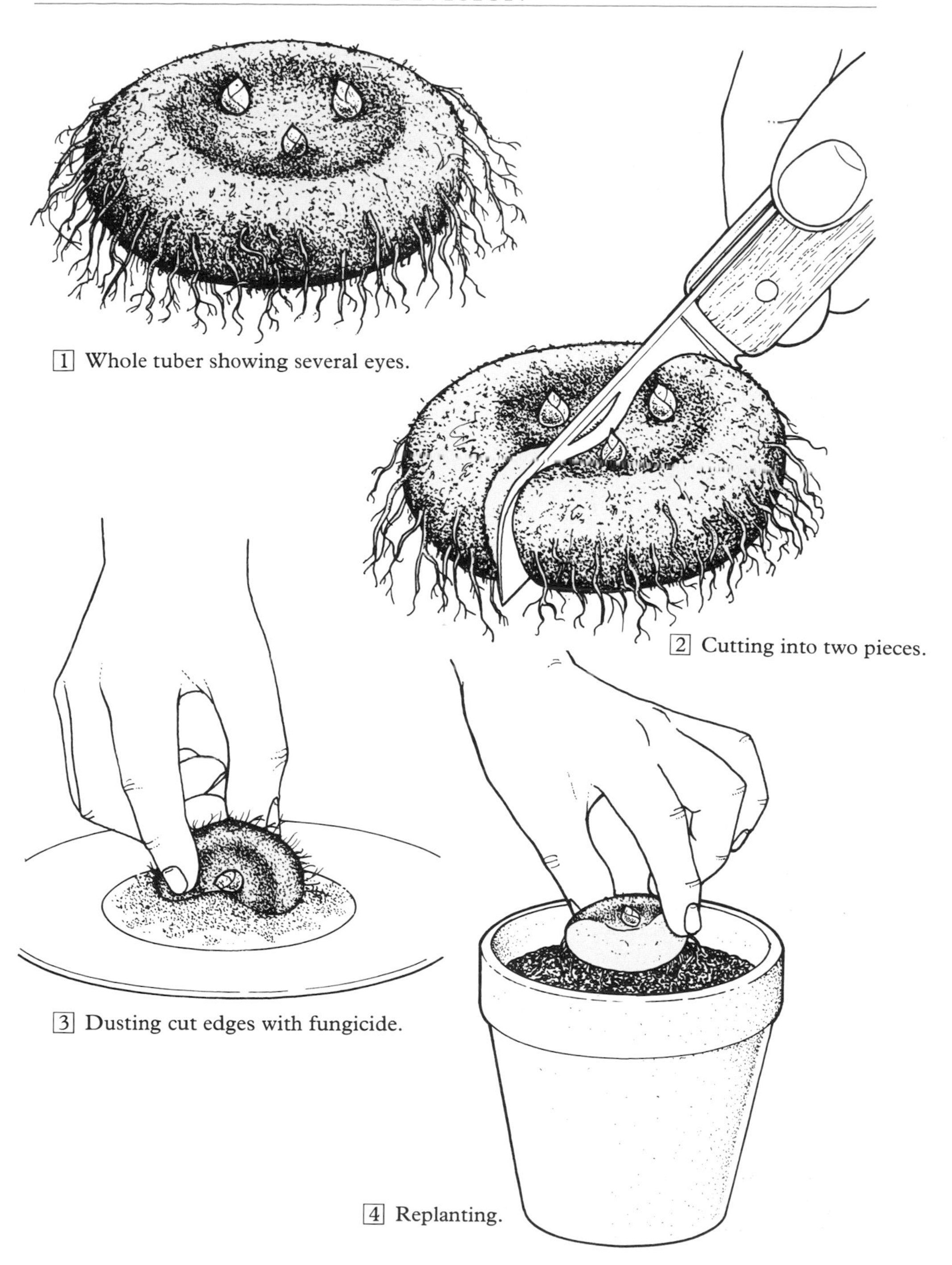

1 Whole tuber showing several eyes.

2 Cutting into two pieces.

3 Dusting cut edges with fungicide.

4 Replanting.

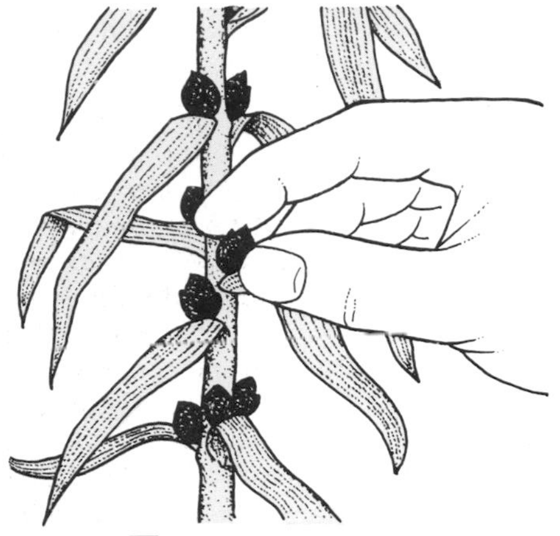

[1] Collecting bulbils.

[2] Putting bulbils in pot.

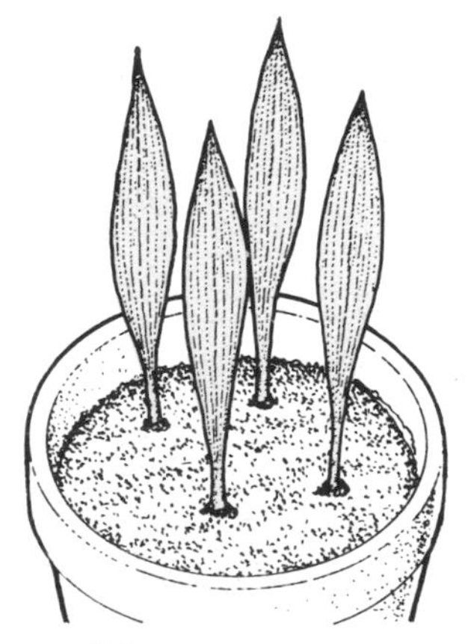

[3] New shoots.

Bulbils

[1] A number of bulbous plants produce tiny bulbs in the leaf axils, up the stem. These are called bulbils and are produced in quantity by some plants; naturally they fall to the ground as they mature after the plants flower. To propagate your plants you need to pick the bulbils off before they fall. Once they are ripe they will come away from the stem quite easily. Not all the bulbils on the stem mature at the same time, so you will need to go over your plant several times.

Plants that don't normally produce bulbils in quantity can be induced to be more generous if you can bring yourself to pick off all the flower buds as soon as they form.

[2] A loam-based potting compost is suitable for the bulbils. Set them in the pot rather as if they were large seeds. They will stay in their pot of compost for a year, so don't set them too close – put about four in a $3\frac{1}{2}$ in (9 cm) pot. Dib them in so they are completely covered; water the pot, label it with the date and name of the plant, and put it in a cold frame.

[3] Keep the bulbils growing on steadily for 12 months. When the young shoots begin to die down, knock the plantlets out of the pots and transplant them to nursery beds to reach flowering size. Take care to keep them weed free, and don't let them go short of water while they are growing.

Bulb offsets and cormlets

Many bulbs increase themselves very efficiently by forming small, 'junior' bulbs around the parent. These are called offsets, and can be separated from the main bulb and grown on very successfully.

[1] Daffodils increase quite quickly, forming large clumps after a number of years. Removing the offsets and re-planting the clump helps to keep it healthy and flowering well.

Lift the bulbs in summer, as the foliage is dying down. Be careful how you insert the fork – it's very easy to spear the bulbs and ruin them. Prise them carefully out of the ground and shake off any loose soil.

[1] Lifting group of bulbs with fork.

[2] The offsets are easily visible at the side of the parent. They are smaller, and are attached to the parent at the basal plate. When broken away, they can be planted and will produce a flower the next summer; the following year they will produce another offset themselves. This first offset is enclosed in the same scaly covering or tunic so that the bulb appears to be a twin – you can often buy these 'double nosed' bulbs in garden shops. The year after, the offsets become more separate until they are only attached at the base. This is the stage at which they can be separated.

Some bulbs (such as tulips) fade away after flowering, so that when you dig the clump up, only the small, young bulbs are left. These can be treated in the same way as the separated offsets.

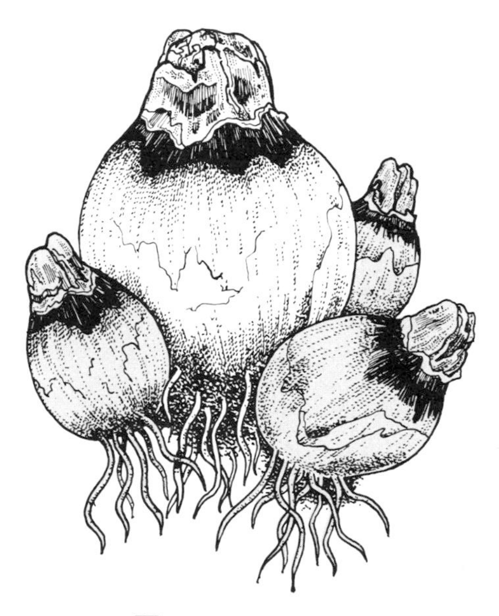

[2] Bulb with offsets.

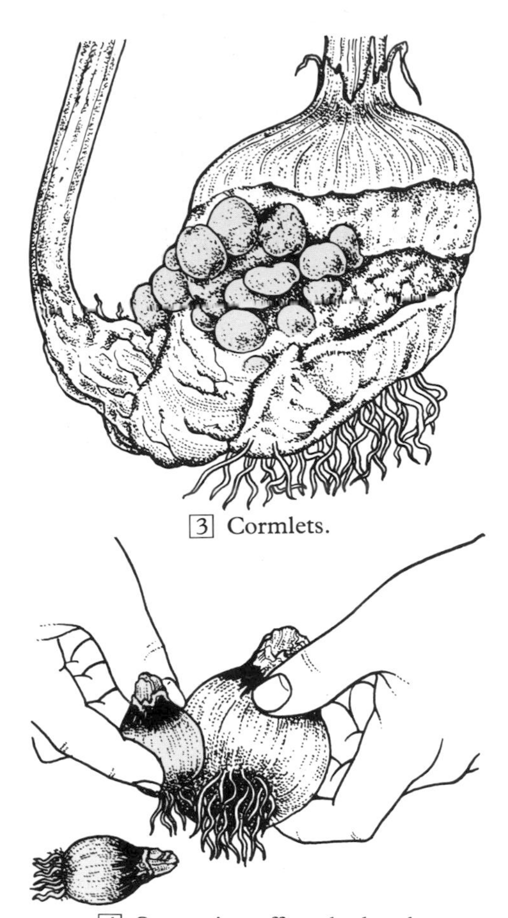

[3] Cormlets.

[4] Separating offsets by hand.

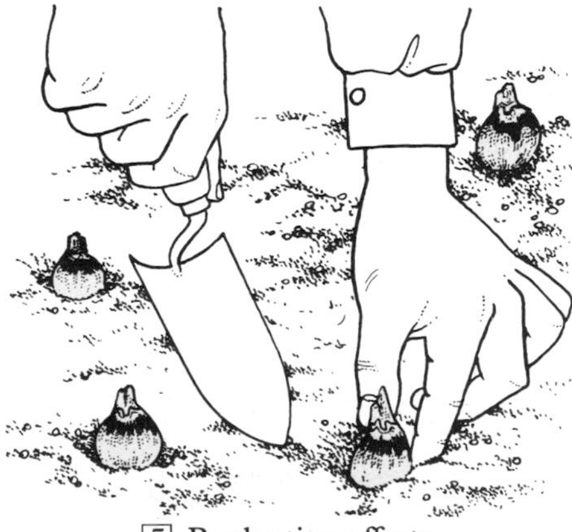

[5] Replanting offsets.

[3] Corms, such as gladioli and crocus, are slightly different to bulbs, and produce offsets in a different way. The original corm, once planted, produces roots and a shoot. As the flower forms, a new corm begins to develop above the old one, which shrivels away. The new corm grows and develops, storing food for the following year; it also produces a number of tiny cormlets round its base. There can be a large number of these cormlets by the end of the growing season.

[4] The bulb offsets and cormlets are easily separated from the parents. Push the offset away from the mother bulb with your thumb until it snaps away at the base.

Gladiolus corms are usually lifted in the autumn and stored in a frost free place over winter. The little cormlets can be pushed off with your thumb at this time. They can be stored in the same conditions as the parents, but as they are so small they dry out more easily. Storing them in slightly moist peat keeps them plump. Next spring, plant them out in a nursery bed and grow them on for two years.

[5] Daffodil offsets can be planted just like newly bought bulbs, directly where you want them to flower. Flower buds should already have been initiated, and the disturbance of replanting shouldn't affect flowering the following year.

Any blind (flowerless) bulbs should be fed with a liquid fertilizer to build them up for flowering the following season. All daffodils benefit from a high potash liquid feed after flowering, while the leaves are growing.

Bulb scales

Bulbs such as lilies are made up of modified leaves which form closely wrapping scales. At the base of each of these scales is a rudimentary bud; this bud can be started into growth and will form a new plant. In this way, one bulb can be completely taken apart to make a large number of plants – or just one or two scales can be used without doing the parent bulb any harm.

[1] Lilies don't take too kindly to being disturbed once they have settled into a garden. There's no need to dig up a bulb to remove a couple of scales (though it is easier to see what you're doing!); you can carefully dig down to the bulb.

Wait until late summer when the plants have finished flowering and dig away the soil at the side of the stem. You can use a trowel or handfork to start with, but as you get nearer to the bulb, scrape the soil away with your fingers to avoid doing any damage. Uncover the side of the bulb completely, right down to its base.

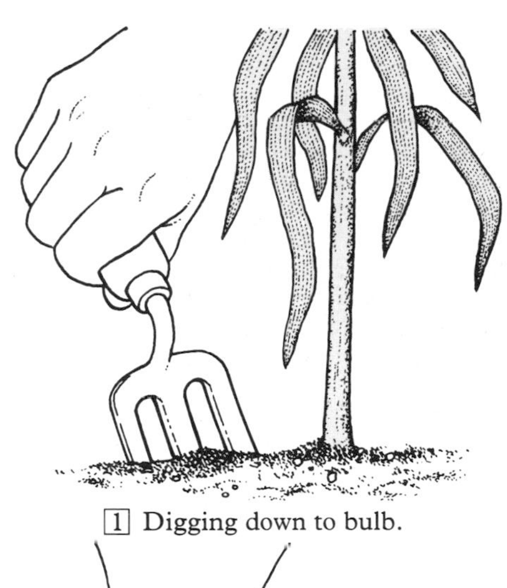

[1] Digging down to bulb.

[2] If a scale is to form a new plant it must be removed entirely; be careful not to snap it above its base. It's quite tricky to do this underground, but out in the daylight it's easy.

Use clean, unblemished scales (not the ones that have been damaged in transit or nibbled by slugs). Pull gently at the tip and the whole scale should snap off cleanly. Replant the parent bulb, which will suffer no ill effects from the removal of two or three scales.

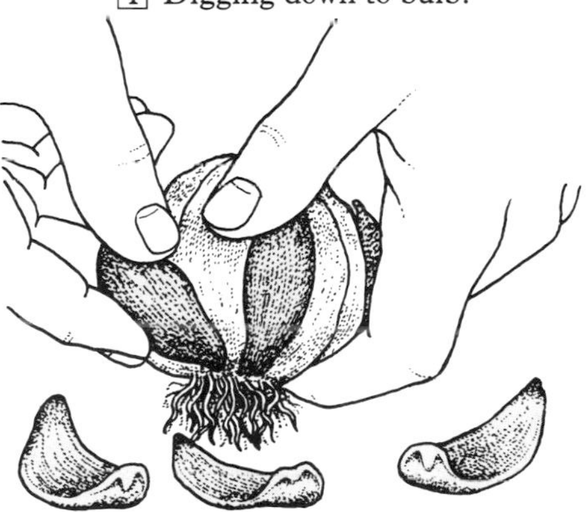

[2] Pulling off scales.

[3] Because the scales have been wounded by being snapped off they can be subject to rotting. Give them a thorough dusting with fungicide powder before going any further.

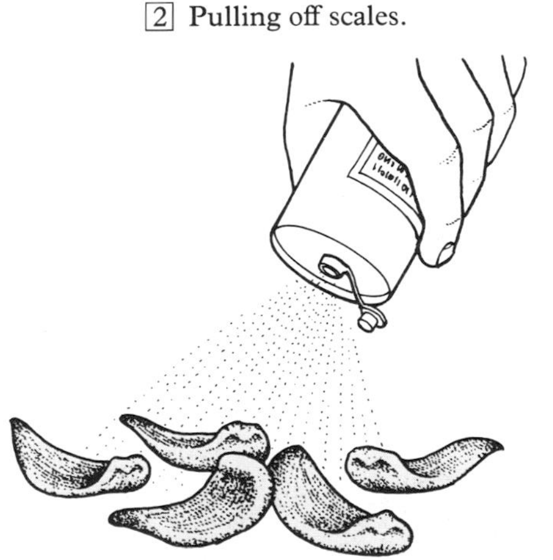

[3] Dusting scales with fungicide.

[4] Scales in plastic bag of compost.

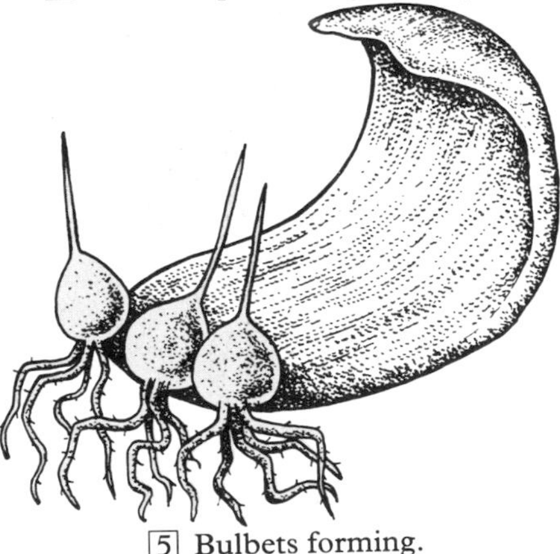

[5] Bulbets forming.

[6] Potting up scale and bulbets.

[4] The scales can be dibbed base first into a pot or tray, but the easiest way to deal with them is to put them into a plastic bag of compost. Use a moist seed and cuttings compost with a little extra sharp sand well mixed in, and half fill a strong polythene bag. Add the prepared scales and stir them lightly into the compost so they are completely covered. Tie the neck of the bag and put it in a warm place such as an airing cupboard.

[5] After a few weeks, tiny bulbs should have formed at the base of the scales, complete with roots and shoots. You will probably first notice little spear-like shoots appearing through the compost: by carefully turning and gently shaking the bag occasionally you will be able to bring some scales to the surface to see how they are progressing. Blow a little air in through the neck if the bag collapses on to the compost. The time taken for the bulblets to form – usually a couple of months – depends on the conditions in which the bag is being kept.

[6] Once new plants have begun to form, the scales should be brought out and potted up. Use a peat or soil-based potting compost and just bury the bulblets – not too deeply. Grow them on in the pots until autumn, when they can be turned out and the plantlets separated. Line them out in nursery beds in the garden and grow them on there for a further year.

Scoring bulbs

[1] Some bulbs don't produce offsets very freely, hyacinths for example. These are not composed of loosely packed scales like the lilies, but are tight and solid, like an onion. The bases of the scales will still produce bulblets if they are exposed.

Cut a cross on the base of the bulb with a sharp knife. The cut should be deep enough to reach the growing point within the bulb, usually about $\frac{1}{4}$ in (5 mm).

[2] Leave the cut bulb in a warm, dry place for the cuts to open out and start to callus; two or three days is usually sufficient. A dusting of fungicide powder will help prevent rotting. After the callusing period, the bulbs should be kept in a slightly moister atmosphere to prevent shrivelling – putting them scored side up in a tray of slightly moist sand in the airing cupboard provides the right sort of atmosphere. It will take 10–12 weeks for bulblets to appear on the cut surfaces.

[3] Once the bulblets are evident, plant the scored bulb upside down in a pot about 2 in (5 cm) deep, so the tips of the bulblets are just covered with compost. Keep the pot in a sheltered place – such as an unheated greenhouse or frame – over winter, and next spring you will be rewarded by the sight of young shoots appearing through the compost. Keep them in the pot until autumn, when the plants can be turned out and separated. They will need to be grown on for two or three years before they flower.

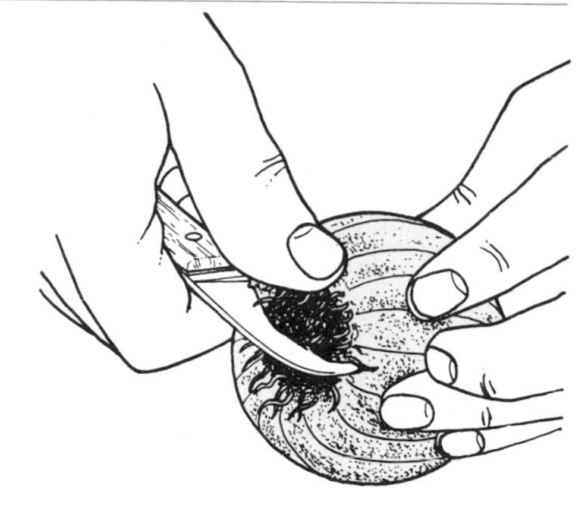

[1] Cutting base of bulb.

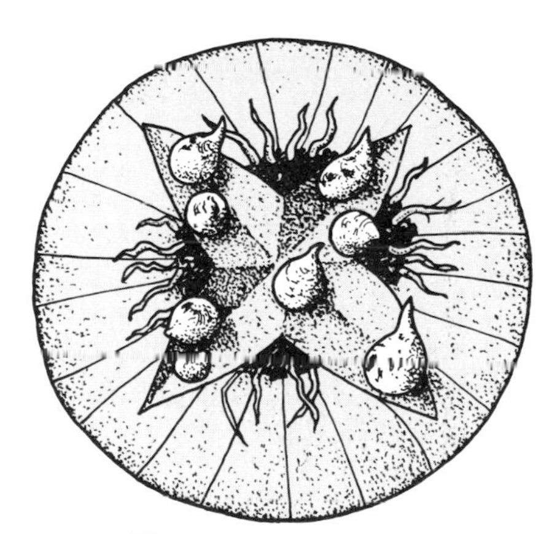

[2] Bulbets forming.

[3] Potting scored bulb and bulbets.

[1] Severing sucker with spade.

[2] Digging up sucker.

Suckers

Suckers are shoots that arise from the roots of certain plants, so that in nature, a small thicket of stems is eventually produced. On many garden plants the suckers are really rather a nuisance, but they're an easy method of propagation. *Rhus typhina* (stags horn sumach) throws up suckers with gay abandon. They are strong and sturdy and 'true to type', i.e. identical to the parent plant. As with root cuttings, make sure the plant you are trying to increase is not grafted on to a different rootstock.

[1] As the sucker is already provided with its own root system, it only needs to be separated from the parent plant. This can be done with a blow from a sharp spade, severing the main connecting root a little way behind the sucker. If you are not sure exactly where to aim the spade, dig away some of the soil between the parent plant and sucker to expose the root so that you can see what you're doing. Some roots are pretty tough, but a pair of loppers should deal with them. Most plants have quite thin, small roots that can be easily cut with a spade.

[2] Once severed, the sucker can be left for a few days to establish itself, or lifted straight away if more convenient. Dig all round the stem with a spade, and lift the young plant with as much soil clinging to its fibrous roots as possible. Transplant it to its new site immediately.

Because some suckers are quite large when they are separated they can be a little more difficult to get established. Water them in to their new position thoroughly, and shelter them from strong sunshine and drying winds for several days.

Offsets

[1] Detaching sempervivum offset.

[1] Offsets are complete new plants which develop alongside the parents and are attached to them. Sometimes they are attached to the base of the crown, sometimes by underground stems or sometimes (like sempervivums and echeverias) by overground stems or short runners. The offsets usually have their own roots and only need detaching from the parent. Offsets on runners should be pulled away complete with the runner stem; if this is left on the parent it will die back into the crown and may cause rotting.

A plant with offsets round its base (such as billbergia) should be tapped out of its pot and the offsets separated and teased apart by hand.

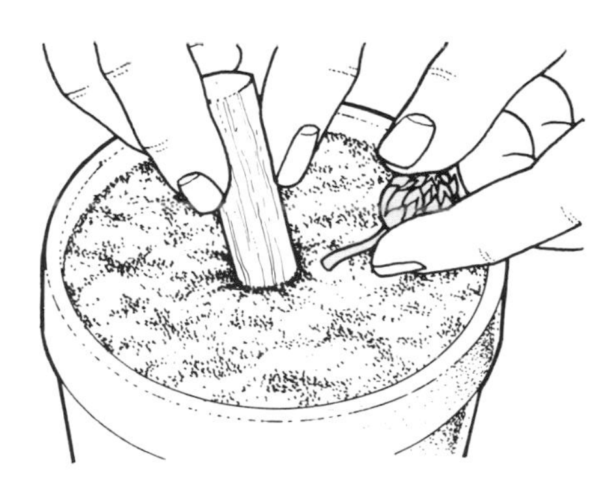

[2] Potting offset.

[2] Offsets which have not yet developed roots should be potted into sandy seed and cuttings compost. Firm them in so their bases are in good contact with the compost. Peat or loam-based potting compost is suitable for offsets that are already equipped with their own roots. Water them in well and keep them shaded from bright sun for a few days.

Some bromeliads die after flowering, but produce numerous offsets before they do so. These should generally be split up and repotted separately if they are to thrive – they don't do well if they are just treated as 'a pot plant' and left alone to jostle for space.

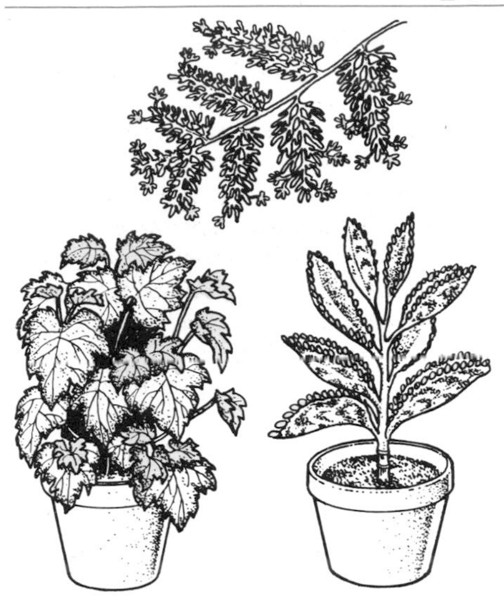

[1] *Asplenium* (top), *Tolmiea* (left) and *Bryophyllum*, showing leaf embryos.

[2] Misting leaf on compost.

[3] New plantlets growing.

Leaf embryos

[1] One of the most fascinating (not to mention the easiest!) methods of propagation is by leaf embryos. These are tiny, baby plants that form naturally, and often in great profusion, on the mother plant's leaves.

Bryophyllum (sometimes called kalanchoe) is one of the most prolific at producing babies – they appear in their dozens round the edge of every leaf. Each plantlet has a pair of leaves and a couple of tiny, hair-like roots: it drops off when the plant is brushed or shaken and takes root in the compost. These plants can then be pricked out into new pots, or the babies can be removed from the leaves before they fall, and laid on the surface of a pot of compost.

[2] Tolmiea and asplenium are not quite so simple. The leaves, complete with embryos, should be removed and laid flat on the surface of a tray of seed and cuttings compost. The atmosphere must be kept humid – spraying with a hand mister is useful. In time, roots will grow into the compost and new plants will get established.

[3] Once the plantlets are growing well, detach them from the mother leaf where necessary and move them into individual pots of peat-based potting compost. Take care not to break the roots, which are often rather fine and fragile. Firm the plants in gently with your fingertips.

What might go wrong

Some of the divided portions die after replanting

•Some damage to the plant is inevitable during division, and if severe enough, the plants won't survive. Tease them apart as carefully as possible, trying to tear them along their natural lines of weakness. •Autumn divisions face a cold, wet season immediately ahead. In heavy soils, particularly, spring is usually a more successful time. •In a dry spring, new plants may require watering to help them establish. •Make sure each division has plenty of fibrous roots. •Don't be too greedy and divide the plants into very tiny portions.

The plant is very large and almost impossible to split up

•It pays to divide all border plants every three years or so, even if you don't want to increase your stock. Neglected clumps can be very tough to deal with. A few well-aimed blows with a sharp spade should cope with the worst customers.

Divisions and suckers are different from the parent plants.

•Before digging up suckers, check that the parent is not grafted on to a different rootstock. •Some variegated plants tend to revert to the plain-leaved form very easily. Rogue out any non-variegated divisions as soon as visible.

Divided root tubers produce no growth

•Root tubers like dahlias must have a portion of bud-bearing stem included with each portion.

Bulbs, corms and tubers rot when split, scored or scaled

•Fleshy bulbs and tubers are easily damaged and attacked by disease. Make clean cuts and treat them with fungicidal powder. •Don't overwater the compost.

Plants suitable for division

Blueprint method Any plant with a fibrous root system and a multi-stemmed crown. Many alpines and rock garden plants can be divided, including *Armeria*, *Aubrieta*, *Alyssum*, *Campanula*, *Erigeron*, *Gentiana*, *Geranium*, *Primula* and *Saponaria*.

The vast majority of herbaceous border plants are also suitable, among them *Achillea*, *Aster*, *Campanula*, *Centaurea*, *Coreopsis*, *Doronicum*, *Echinops*, *Hosta*, *Papaver*, *Primula*, *Pyrethrum*, *Solidago* and *Veronica*.

Rhizomes *Canna*, bearded *Iris*, *Zantedeschia*.

Tubers *Begonia*, *Caladium*, *Cyclamen*, *Dahlia*, *Gloriosa*, potato, *Ranunculus*, *Sinningia* (gloxinia).

Bulbils Lilies, especially *Lilium tigrinum*, *L. bulbiferum* and *L. sargentiae*. *Lilium candidum*, *L. chalcedonicum* and *L. testaceum* will produce bulbils if the flower buds are removed.

Begonia evansiana produces tubercles in the leaf axils, which are treated in exactly the same way as bulbils. The Egyptian tree onion, *Allium cepa aggregatum*, produces bulbils in the flower head.

Bulb offsets *Allium*, *Amaryllis belladonna*, *Camassia*, *Crinum*, *Endymion*, *Fritillaria*, *Iris*, *Leucojum*, *Muscari*, *Narcissus*, *Oxalis*, *Polianthes*, *Scilla*, *Sprekelia*, *Tulipa*, *Vallota*.

Cormlets *Acidanthera*, *Crocus*, *Colchicums*, *Freesia*, *Gladiolus*, *Ixia*, *Sparaxis*, *Tigridia*.

Bulb scales Fritillarias, lilies.

Bulb scoring *Hyacinth*, *Lachenalia*, *Scilla*.

Suckers *Ailanthus*, *Chaenomeles*, *Corylus*, *Daphne genkwa*, *Jasminum*, *Malus*, *Nandina*, *Passiflora*, *Philadelphus*, *Prunus*, *Rhus typhina*, *Robinia*, *Rosa*, *Symphoricarpos*, *Syringa*.

Offsets *Agave*, *Billbergia*, *Echeveria*, *Lewisia*, *Sempervivum*, *Vriesia*.

Leaf embryos *Asplenium bulbiferum*, *Bryophyllum daigremontianum*, *B. pinnata*, *Tolmiea menziesii*.

GLOSSARY

Annual A plant which grows from seed, flowers and dies within one year.
Axillary Arising from the leaf axil (where the leaf joins the stem).
Biennial Grows from seed one year, overwinters, and flowers and dies the following year.
Bleeding Loss of plant sap from a wound.
Bottom heat Artificial warmth provided below the rooting medium.
Broadcasting Scattering seed randomly over an area.
Bud grafting Budding.
Calcifuge Plants which dislike lime in the soil.
Cambium Living cells found between the rind (bark) and the pith of plants.
Chitting Starting into growth before planting (seeds, tubers etc).
Cotyledon Seed leaf.
Cultivar Cultivated variety.
Damping off A fungal disease of seedlings, causing the stems to collapse at soil level.
Dibber A tool for making holes in the soil to receive plants.
Dormancy A state in which plants or seeds are alive but not in active growth.
Embryo The living plant within the seed.
Hardening off Getting plants gradually acclimatized to outdoor conditions.
Hardy Able to survive outdoors all through the year.
Heel A small tail of older wood taken with a shoot used as a cutting.
Heeling in Temporarily covering a plant's base or roots with soil.
Hybrid A plant produced by crossing two different varieties.
Lateral Side shoot.
Light A framed sheet of glass or plastic on a garden frame.
Maiden A one year old tree.
Mulch To cover the soil with material to retain moisture and suppress weed growth.
Node The joint where a leaf meets the stem.
Pan A shallow, round container or pot.
Perennial A plant surviving for several years.
Petiole Leaf stalk.
pH A measure of acidity or alkalinity of soil.
Pricking out Planting seedlings in fresh compost, at a wider spacing.
Scion A shoot or bud grafted on to a rootstock.
Sport A genetic mutation giving rise to a different type of plant or flower.
Stock Rootstock: or a plant kept specially for propagating from.
Stool Roots of a plant kept for producing cuttings, especially chrysanthemum.
Stopping Removing a plant's growth point to encourage branching.
Systemic Carried through the whole plant (especially insecticides).
Transpiration The continual loss of moisture through a plant's leaves and its replacement from the soil.
Tubercle Small rounded prominence which may be borne in leaf axils (e.g. Begonia) or on roots (e.g. legumes).
Variegated Leaves marked with some colour other than green – usually yellow or white.
Vegetative propagation any method other than by seed.
Whip Another term for a maiden.

INDEX